DATE DUE

MAY - 2 2000			

Parfit
  Last stand at
Rosebud Creek

Last
Stand at
Rosebud
Creek

Last Stand at Rosebud Creek

Coal, Power, and People

Michael Parfit

E. P. Dutton   New York

For information contact: Elsevier-Dutton Publishing Co., Inc., 2 Park Avenue, New York, N.Y. 10016

Library of Congress Cataloging in Publication Data
Parfit, Michael.
Last stand at Rosebud Creek.
1. Electric utilities—Montana—Rosebud County.
2. Electric power-plants—Environmental aspects—Montana—Rosebud County. 3. Strip mining—Environmental aspects—Montana—Rosebud County. 4. Environmental policy—Montana—Rosebud County—Citizen participation.
I. Title.
HD9685.U6M96   333.79′3215   79-27264

ISBN: 0-525-14357-2

Published simultaneously in Canada by Clarke, Irwin & Company Limited, Toronto and Vancouver

Designed by Mary Beth Bosco

10 9 8 7 6 5 4 3 2 1

First Edition

To
Professor Ray Canton
and
Deborah N. Parfit

Introduction

On February 29, 1980, three Wisconsin power companies abandoned plans to construct a nuclear power plant near Sheboygan, Wisconsin. Instead, they said, they would produce new electricity by burning coal.

The decision was part of a trend. Nuclear power development, once the great hope of the century, has almost come to a halt, stalled by accidents, regulatory difficulties, and widespread public mistrust. Between January 1, 1979 and the time of the Wisconsin decision, utilities across America had already canceled plans for fifty-one nuclear generation units which would have produced a total of 52,123 megawatts, or enough power for the domestic needs of about 35 million people.

In the place of nuclear power, as the Wisconsin decision indicated, the production of electricity with steam heated by burning coal is booming. One hundred and sixty two new coal-fired power plants are scheduled to start pumping out electricity between 1979 and 1993. Their total output will be about 89,440 megawatts, or enough for 59 million people. Anyone who reads this book will at one time or another use this electricity, perhaps even to illuminate these pages.

So it seems that the horror of unknown infiltrations of radiation and uncertain odds of catastrophe are being replaced by an old-fashioned kind of fire. And the question, surely made more urgent by what is now known about the relationship between the burning of coal and its effect on human health and well-being, is whether the replacement is really as benign as the nation so fervently hopes.

This is not a book of polemics. It is a simple story of eighteen people of various professions and persuasions who, by choice or accident, became involved with two units of those 162 new coal-fired power plants. But it may offer part of an answer to that question. The rest of the answer may only be found in the sky and the rain and the precious waters of America over the next fifty years.

This is a book of nonfiction. The events described within are based on the truth as perceived by their participants. No names have been changed.

MICHAEL PARFIT

Last
Stand at
Rosebud
Creek

Prologue

1

... and those people who could not fit into the small courtroom spilled out onto the foyers of the second and third floors of the courthouse, along with uniformed officers of the Rosebud County Sheriff's Department who were on hand to guard against possible violence. ... Throughout the night the testimony taken inside was punctuated by a rumble of conversation in the corridors of the building and the honks and squeaks of the coal trains which from time to time rumbled along the Burlington Northern tracks south of the open windows.

—*Forsyth Independent*
December 25, 1974

There was a snowstorm that night. It blew down from Canada in racing sheets of white, and it smothered Rosebud County. Cattle steamed in the fields and clustered together for warmth. Blown snow packed hard on the windward sides of trees, telephone poles, and highway signs and collected in the crevices of the cliffs of red stone the ranchers call scoria. Snow rattled in the dead leaves of the trees outside the county courthouse in Forsyth, hissed along the highways like clouds of bitter dust, and smeared the roads with ice.

But the people came. They fought the storm and the darkness to drive in from Hardin, Lame Deer, Colstrip, Birney, Miles City, Busby, Ashland, and Billings, from all over eastern Montana. They parked their cars and pickup trucks on the streets, wrapped their coats around them, and poured into the courthouse and up the marble stairs. By 7:30 P.M. the courtroom was full and there were people crowded at the big oak doors.

Inside the high room all seventy-five spectator seats were filled. The fourteen padded jury box seats had been filled for almost an hour.

3

There was talk of moving the meeting to the Forsyth Library basement, a couple of blocks away, where there might be more room, but the people in the jury seats refused to move. Men sat on and under the steel-topped tables normally reserved for prosecution and defense lawyers. Men and women stood in the back. There were people outside on the stairs. Only the chair up on the dais under a mural of Moses bearing tablets from God's fire, the leather-covered chair where the judge sits, was empty.

The heat in the building was turned up full. Beads of wetness gathered on the window glass and on the faces in the crowd. The radiators sizzled. Someone opened a window. A gust blew the curtains and slapped a few faces with cold but it offered no relief, only a sweaty chill. Beyond the doors a chanting began: "We want in! We want in!" Outside, a coal train clattered past, picking up momentum on the old tracks, carrying its cargo east across the prairies into the power-hungry cities of the Midwest.

In the courtroom a young man named Mike Moon stood nervously before the crowd. He was slight of figure and just out of college; he looked too much of a boy to run this kind of crowd. That morning he had dressed carefully: a western shirt with snap pockets for the ranchers, a tie for the power company executives, and Vibram-soled boots for the environmentalists. These tokens did not help his confidence. Now he explained the purpose of the meeting. It had been called, he said, as part of a series of meetings to gather input on a draft environment impact statement covering the construction of two 750-megawatt coal-fired power plants at the town of Colstrip, thirty-five miles south of Forsyth. *That is what this meeting is about. That is all this meeting is about. Please confine your remarks to this subject.*

The faces of the crowd took his words impassively, digested them without remark, and gazed back. Moon paused, uneasy. Strange that so many people should come here through the storm to fight about a power plant. Not odd when you considered what he had been doing the past month—running similar meetings across the state—but strange in general. The plant itself stood before him, over the heads, like a ghost, a crazy Flying Dutchman called Colstrip Units Three and Four, built into substance out of the hundreds of slides, schematics, and charts he had shown a dozen audiences like this one over the past weeks. It roared and thundered and poured its plume of steam east with the wind, a ship careening on the turbulent overburden of eastern

Montana, carrying its cargo of fear or riches west to an unknown shore. Moon looked out over the audience. The chanting had subsided in the hall; there was a stirring by the door. A pair of Rosebud County Sheriff's Department officers shouldered their way into the room. They stood in the back. Their faces were blank.

Moon shuffled a stack of three- by five-inch cards on which those who wanted to speak had written their names and addresses. He wondered how many of them had really thought it out. This was their moment of decision: a public hearing may not have much effect in the halls of legislation, but it uncompromisingly defines the motives and emotions of a community and, even more clearly, the beliefs of the individual. You speak; you make yourself known; you are no longer a bystander in conflict. It is a bit like publishing poetry—you forgo small talk and reticence when your passion is spread-eagled on the page. All up and down Rosebud County, at the Joseph Café in Forsyth or the B & R Bar on the edge of Colstrip, once these people had spoken they would have no neutral ground.

Moon glanced through the names. Some he knew; they had long ago made their decision and their enemies. Some were to speak for the first time. Would it be worth it? The list was like a cast of players:

MARTIN WHITE. Manager of the town of Colstrip. Fair hair, square jaw, forthright blue eyes. Company man, loyal beyond belief. White sometimes stood up and said the strangest things.

MYRON BRIEN. A Northern Cheyenne Indian. Usually the only member of his tribe at a meeting who spoke in favor of the power plant. There were rumors he was a hired gun, paid to speak. Already he was getting well known on the circuit for his earthy eloquence.

BROTHER TED CRAMER. A monk. The man in the roman collar who was sitting in the back. This week's representative from God.

PATRICIA MCRAE. English teacher. One of the McRae clan. In eastern Montana the name carried, coulee to sandstone ridge, like the bellow of a Hereford bull. The McRaes were hard to stop when they got to talking, but usually they were worth hearing, too.

MARIE SANCHEZ. Northern Cheyenne. Moon had heard that the American Indian Movement would be represented here tonight. This must be the one.

LOIS OLMSTEAD. Homemaker, Colstrip. That must be the woman in the back, wearing a skirt. Wearing a skirt!

EVAN MCRAE. Another bearer of the familiar name; this, too, was a face that had been to many meetings: sincere, balding, wary, calm eyes. Down on the Rosebud they call him Duke.

BILL PARKER. Another Indian. Probably a half-breed. Incredibly verbose, or so it seemed when he spoke, pouring out the testimony. There was a texture to his language that echoed like eloquence in passing, but it all went by too fast.

DON BAILEY. Another Rosebud Creek rancher. Sometimes visibly hostile. But those ranchers—you had to hand it to them. Organization. They knew what they were doing. Even though sometimes they made the whole exercise seem too much like a game.

PATRICK HAYWORTH; MIKE HAYWORTH. Two new names. Brothers, no doubt, with first names like that. They would, of course, be on the same side of the issue. But which side?

CLIFFORD POWELL. Unknown. Hadn't spoken before. The card said he was from Billings. Probably someone down in that clump of construction workers. A few weeks ago one of those people had been totally drunk and ready to fight even the microphone; Moon hoped that wouldn't happen here.

SHEILA MCRAE. The high school girl sitting next to Duke, probably. She was looking across the room into the group of workers. She had a look that was almost like despair.

CHARLIE WALLACE; BARBARA WALLACE. New names: Husband and wife? What were they getting themselves into tonight? Suppose they came out on opposite sides of the issue; wouldn't that make things interesting!

TOM WIMER. Another Rosebud rancher. Unfamiliar, though. He looked like the type that would show up only once or twice, nagged into it by his activist neighbors.

WALLY MCRAE. Moon looked around the room. There he was, hunched under a table near the front, the most familiar of the clan, of all the ranchers, the celebrity of the coal wars. He was talking—no, listening!—to a younger man with a sharp-edged face who appeared to be one of the construction breed: longer hair, tan right up to the roots of it. McRae's eyes were narrow glitters in an impassive face, like the slits cut in a leather mask. Last night in Hardin, McRae had nearly got in a fight with a construction worker right there in front of the audience; at least that's what Moon thought was going to happen. It had been uncharacteristic of him, almost to an extreme, but that was the way these things were going.

Moon looked out at the crowd. It was quiet at last. The coal train had gone. The people were waiting. There was a smell of beer, a smell of old wood and plaster, and too much steam heat. The radiators ticked. The two officers shifted their stances. Neither of them was chewing gum. Moon slid the last card back into the deck and shuffled. They were bits of a puzzle, these people, separate little chunks of a thing that was happening that even he did not fully understand: Put them together, listen to them talk, give them time to weep and pray and shout and argue and giggle and grow silent with anger, and perhaps you would come up with a small approximation of truth.

He shuffled again, then picked off the top name, announced it, then moved to the side of the room to let the speeches begin. By the time the evening was over those names and faces would mean more to him than they did now, and maybe he would learn something and maybe he wouldn't. He just hoped that the night would not end in violence.

2

On a morning in December I climb to a place on the spoil pile where a child has dug out a shallow cup in the slope and laid a board in it, and I sit on the board and look out at the power plant. I have come here to exercise my fear, to promenade my anxiety up here where I can see its source filling the air with cloud.

The power plant dominates the valley below me. It occupies a space that could be used to conduct two simultaneous football games before one hundred thousand screaming fans. Its foundations are forty feet deep. Its superstructure is over two hundred feet high and is made up of salmon and green corrugated metal, silver pipes, and reddish steel girders. It looks like an unfinished gantry, as if the stacks that stand behind it blasting steam into the cold sky were rockets instead of vents and were poised to fling themselves, not debris and moisture, into the heavens.

From where I sit today, and in fact, from almost everywhere in Colstrip, the plant is never entirely visible. It hides behind the cloud of its own making, its most visible product, the water vapor that gives it its colloquial name, the steam plant. West and north of it are two huge banks of cooling towers: Each bank is a row of squat nozzles pointed at the sky, from which pour sheets of gray cloud edged silver by the sun. The clouds sweep across the plant like hurrying patches of fog, and join their amorphous curtains with the dozens of squirts and hisses of steam that gush from the plant itself as if it were a kettle badly wounded. The stacks, each five hundred feet tall, reach up from the maelstrom of steel and steam, but they too make weather. Each emits a dense column of water vapor that bends down with the wind like a flag and races southeast toward the morning sun, casting hurtling shadows across the plant itself and the dry land beyond. These twin columns of effluent lie together across the sky; in the turbulence created by the thrust of the stacks up into the wind, they finally twist together two miles downstream and braid themselves into a long, flat cloud that gently dissipates over the ranches along Rosebud Creek and leaves only a faint red stain against the horizon.

The plant bellows, whines, clatters, groans, screams, thunders, and trumpets. The sound seems to batter me here, half a mile away, and I know that it seems almost as loud up at an old homestead I have visited four miles north on a dirt road. The dominant theme of this noise is a constant hissing roar, like the infinite exhalation of a fumarole. Against this background the plant creates subtle diversions: the occasional sound of a clatter, as if a pickup truck full of tools had fallen downstairs; a slender whine; a grinding as if of unmeshed gears; the tight slam of colliding locomotives; and an occasional full-throated baritone shriek. Above all of these noises comes the regular interruption of the plant's own internal voice calling out on loudspeakers across the tumult. ARAXNID BROZZA LINE THREE! GARZNAVAD RD ZAG, XRVARD, LINE TWO! AAAZD FRAXT ZINE POUR! The voice carries even farther out over the land than does the general roar, reaching across the tall brown grass up into the coulees, bouncing echoes off the red stone cliffs of the little mesas where the pine trees grow. The power plant is calling names.

Three days ago I came home from Helena. Home is now Colstrip, a 10- by 38-foot trailer parked on the scoria slope of Burtco Court east of town. My trailer is tied to Colstrip by three umbilical rubber hoses: water, power, sewer. I'm tied here by the power plant. Like everyone else here on this bald and stony hillside, bounded on one side by old spoils and on the other by power lines, I am employed by it. Indirectly, of course; I am not summoned each morning to its interior, and I do not draw a paycheck from Montana Power Company. But if it were not here, neither would I be. I have come to Colstrip, to the shadow of the power plant, to tell its story.

The plant steams before me, at its left side the wide space on the land prepared for its huge neighbors, Colstrip Three and Four, which are so far just as tangible but insubstantial as the steam of One and Two. And as I vibrate to the roar, I am touched by that familiar fear. I am seldom afraid of the infamous blank sheet of paper in the typewriter, a thing that seems, from the literature, to haunt other writers. What scares me is the blank story, the big open space right at the beginning of a piece, the hurtling sphere of reality that surrounds you yet cannot be stopped and examined and picked over and held. Now, right at the beginning of the story, at the point just before I first make contact with the event, the people, the place that is to be its heart, the fear comes, broad and undeniable: the knowledge of inadequacy, of the very vague relationship that the printed word has to truth (although it is quite easy to get around this by arguing that there is no such thing as truth); a fear both relaxed and sharpened by an overwhelming understanding that indeed, something *is* happening here, something wild and important, as endlessly fascinating and terrible as it is impossible to catch. I have a list in my trailer that I brought from Helena, along with a copy of a tape recording. The list contains the names of seventeen people who spoke at a public meeting in Forsyth two years ago; the tape is of their speeches that night. My intention is to talk to each of them, to let them put structure and meaning into the rage of conflict that surrounds the power plant. Eighteen voices, eighteen people, touched, bruised, controlled, or ruled by this machine. I will tie the voices together and let their story tell itself. My intention is good, my goals are admirable, surely? I am sincere, at least. But my fear billows up like the steam. I am about to throw myself into the middle of someone else's turmoil, and all I can hope for is a quick perception, so that the thing that is important in the melee, that is invisible to me now, will not escape my notice in the end.

I leave the plank. I skid down the slope of the old spoil pile, which is rutted and loose like the hide of a dead elephant. Near the bottom children are playing in the corn snow left in the shadows from last week's storm. Their voices are thin under the dominant noise. As I descend the plant fills the horizon and everything beyond it dwindles—the hills, the pine trees, the blue ridges down by the Northern Cheyenne Indian Reservation. The world fills up with its clamor; its own clouds cover the sun. And I know that what really scares me is the suspicion that I'm not the one in control—I'm not the one tying this story together at all, that maybe I'm just character number nineteen.

But the story does not begin today in the power plant's shadow: It starts fifty-two years ago in Minneapolis, Minnesota.

Part 1

3

Engineering science today literally has begun peeling the surface of earth from 20,000,000,000 tons of semi-bituminous coal and is turning the product from openpit mines here to power generation which, in the vision of engineers, ultimately may mean shifting the nation's industrial center to the open spaces where cactus and sagebrush now abound.

—Fred W. DeGuire
June 21, 1925

It was Sunday, June 21, 1925. The great black rotary presses at the *Minneapolis Journal*'s office downtown had already cooled by the time a dapper little public relations man for the Northern Pacific Railroad went outside to get his paper. Glancing with little interest at the front page—EUROPE LINES UP TO PAY DEBTS TO U.S., 1,000 KILLED WHEN REDS INCITE REVOLT IN CHILE PROVINCE, COOLIDGE STUDIES PLANS TO TIGHTEN DRY LAW OVER U.S., BOW AND ARROW HUNTER KILLS ENRAGED LEOPARD WITH BARE HANDS—he turned to the editorial section and smiled.

There, filling the whole of page five with thin gray type and photographs, was his story. Looked good, if a bit overenthusiastic. He read it, nodding at the familiar words, grimacing at a typo, then, satisfied, clipped the page carefully with a razor blade he kept for the purpose, and turned to the financial section. Once again he had done his job well. The company would be pleased. The impact of what he had written drifted cleanly from his mind.

Ignored on his desk, the story continued to shout its strident message of promise, as it was shouting it into households all over Minneapolis. There, too, it would be quickly forgotten. Few people would take it seriously for more than it was: a pebble of speculation hurled with little hope into the sky of public awareness, soon to disappear. Everyone was making similar promises those days, if only to drive up prices of stock; some, like this one, were so preposterous that it was hard to believe they'd ever happen. Montana, industrial center of a nation? People were still trying to homestead out there, and failing.

COAL STRIPPED FROM EARTH'S SURFACE, the main headline on DeGuire's story read, TURNED INTO POWER IN GIGANTIC NEW ENGINEERING ENTERPRISE IN MONTANA.

Below this two smaller, inverted-pyramid-style headlines told more of the story and added commentary of their own:

POSSIBILITY OF SHIFT OF NATION'S IN-
DUSTRIAL CENTERS TO "GREAT OPEN
SPACES" WHERE NOW THE CACTUS
AND THE SAGEBRUSH ABOUND,
SEEN IN HUGE ACTIVITIES
OF THE NORTHERN PACIFIC
RAILWAY COMPANY
AT COLSTRIP

Opposite this, separated by a photograph of a dragline, was:

REVOLUTION IN RAILWAY OPERATION
AND INDUSTRIAL POWER EXPECTED
FROM GREAT NEW ADAPTATION OF
COAL MINING TO MODERN NEEDS
CLOSE TO SOURCE OF FUEL FOR
TRANSFORMATION TO POWER
AND LIGHTING CURRENT
LINES

DeGuire was a competent journalist and a better PR man. If he was well known at all, it was as "the slickest little handshaker east of the Missouri." Years later his widow paid him a simple compliment that would probably do justice to most of us: "He went along pretty well with what he was supposed to go along with," she said, "whether he liked it or not."

His job was to make the railroad look good, and in this story he not only made it look magnificent but cleverly pictured it at the front of a wave of industrial expansion. First, capturing his readers with graphic description, he detailed the size of the mine equipment: a dragline "which would make an excavation for a home with a few scoops and in a few minutes, taking the dirt a block and a half away," a coal shovel "which lifts enough coal at a time to supply the normal family for a season." Then he made his predictions.

"No greater possibilities for the Northwest exist today than in the strip-mining operations which the Northern Pacific Railroad has undertaken here," the story said. "The problem for industrial expansion always has been fuel. So engineers dream of an eastern Montana, the hub of American industry, probably within the day of the present generation and perhaps within a few years."

Later in the story DeGuire tempered the naked optimism of his words by quoting a man named Charles Donnelly, president of the Northern Pacific Railroad.

"No one today," Donnelly said, "could forecast the future of this development accurately."

DeGuire had no way of knowing, of course, that what he had seen out across fifty years of the empty spaces of eastern Montana was the silhouette of the power plant.

4

With a burst of bitter smoke, the hot iron bit down through red hair to flesh and left its mark: a red, stinking welt on the ribs. The iron was removed and another one replaced it fresh from the fire, and a second welt was laid near the first, on the left hip, with another flash of smoke and another scream. As the iron was removed, a man in Levi's and a broad-brimmed hat reached down and with a swift jerk cut away the bull calf's testicles and threw them in a bucket. A small boy, working very seriously, dumped a spoonful of salt in the new steer's bleeding cut, careful to turn his hand to the left instead of the right, and trying not to think of quick, sharp hooves.

The calf bawled again, joining its voice, hoarse and desperate for the first time in its life, to a chorus of groans and screams and bellows that rose with the smoke like a dirge, the lament of the Herefords in the world of man.

As the prone calf was vaccinated, the cowboy with the knife grabbed its left ear and sliced off its tip. He tossed this limp and bloody token in a pile near the fence. The boy behind him with the salt, the "powder monkey," watched. He was about seven years old, he had a narrow, serious face, and his quick eyes glittered, grabbing experience and turning it unconsciously to instinct to be called upon without thought in future years. He knew why the horns of the cattle were removed; he knew what the castrating was for; he was fascinated by the designs of the brand and the smell of the smoke; and he knew that at the end of the day all those bits of ears would be the only way of knowing how many calves had been branded. The little pile would tell the rancher, as precisely as possible, how his year had been so far. The boy was Wallace McRae, third generation on the Rosebud, and the year was 1944.

16

The corral was located on a wide, flat bench in a part of eastern Montana called Rosebud County, in a part of that 5,600-square-mile county that is about a hundred miles east of Billings and about forty miles south of Forsyth on the gravel road. From here the escarpment of the Rocky Mountains lay just over the horizon to the west, and the country was neither flat nor mountainous. It was choppy—broken into a series of small ridges and valleys, like a sea swell sharpening over a reef on the way to a shore of cliffs and broken rocks. Each little valley was a sweep of grass and a gleam of water, and each ridge cut away into a talus of the pink rock that the ranchers called scoria but which is, in fact, red clinker, burned over the centuries by underground fire.

McRae was old enough now to have a sense of country. He knew the waters: He had watched the Yellowstone where it flowed east past Forsyth, where, having left its wild mountain exuberance behind it, it meandered with an absentminded and slightly murky pace toward the Missouri, gathering in similarly quiet tributaries as it went. He had waded across the Tongue, to the east of here, which flowed north between low banks where white men and Indians were still carving fields out of thickets of cottonwood trees. And he had grown to love the narrow, dark wriggle of Rosebud Creek, which wandered down from the Northern Cheyenne reservation town of Lame Deer to pass virtually in his backyard, though yards here were measured in square miles.

He knew the land; he had a feel for it, though he couldn't tell you how. He had spent time in it, as much outside as in. For seven springs the same valley had grown rich green and yellow and blue around him with buffalo grass, western wheat grass, blue grama grass, needle and thread grass, mustard and lupine. For seven summers the valley had gone swiftly brown in the heat, like a delicate skin burned to toughness under the sun. For seven autumns the leaves had flared on the stunted, wiry cottonwoods and box elders that lined the creek and had withered and blown away with the northwest wind. For seven winters he had seen the blown snow, the cattle heaving and snorting in the drifts, and the men going out into the hard cold and coming back with faces blistered by the wind. He had explored the little hillsides and found the springs that looked like large green bruises in the almost-perennial brown. He had played in the steep talus slopes where the scoria outcropped in red-pink bluffs of loose rock and he had thrown handfuls of the stuff down the hills and listened to the pieces tinkle like bits of broken glass. He knew the magic of the trees. He knew they made this land special, that without the red cliffs and the pines, it would seem a desert. But it did not seem remarkable to him that in the space of a few hundred feet of elevation the nearly treeless valleys gave way to forested hills, that all the ridges were edged with groves of Ponderosa pines. They were just a part of his life, softening what would otherwise be a rude landscape. From the valleys the trees seemed a distant presence, offering in summer a promise of shade and in winter, shelter.

The trees, the slow creeks, the grass, and the scoria were part of the pattern of his life, but the resonance of his family's relationship to this rugged place went deeper than that. Wally McRae was already aware that his life, through his parents, touched that primitive time of the Indians; his grandfather John B. McRae had come up the trail from Texas in 1882, just four years after a determined band of Northern Cheyennes had fought their way past massacres and white treachery from a reservation in Oklahoma back to their chosen land. And his father, Donald McRae, who today was the man with the knife whom Wally followed, had seen the desperate struggle of the homesteaders. Lured to the arid lands of eastern Montana in the early 1900s by strange theories about irrigation and dry land farming promoted in railroad company literature, the homesteaders had poured into the state. They had put up their tiny log and frame homes on sagebrush-covered knolls and had waited patiently for the climate to change. The ranchers already there watched with more pity than contempt as the drought of 1918–20 dried them out like grapes on a mat, and as they carried their wizened dreams away in baskets. Donald McRae had heard an Independence Day speech in the Castle Rock Community Hall, north of the Rosebud. "I have a dream," the speaker had said. "I see white frame houses and black-and-white milk cows, on green verdant hillsides; patches of corn, spring houses, flowing creeks; roads, schools, churches." When Wally McRae entered first grade, the Castle Rock Community Hall was gone and the man who had spoken so stirringly of the future was a janitor in the Colstrip school.

And McRae had already learned too well of the hard days of the depression, when beef prices couldn't pay for freight to market, and when winters consumed cattle like a voracious white beast, and left only bones for the spring. He would remember all his life what his parents had already told him, and would tell him again: "It was tough in the thirties; it was damn tough in the thirties. The only reason we're still here is that *we* were tough in the thirties."

5

Calves bawled, cows called. Smoke spurted up to writhe in the tumult and blow away. In the corral, Wally McRae waited with the powder while a pack of older boys wrassled a calf twice their weight, grabbing the rope and tail and heaving until, beleaguered by boys, the calf fell. The smoke puffed. The knife flashed. Legs pinned, the calf humped on the ground like a seal. Wally applied the salt.

Outside the corral, women brought lunch: slabs of bread and meat, apple and cherry pies, and cookies. A young cowboy rode a horse in too fast, and Donald McRae said to another cowboy, consciously within the hearing of his son: "Hadn't ought to run your horse home. Makes him crazy, wants to run to the barn every time." And the child remembered. McRae remembered that lesson—reinforced regularly by his father—so well that the first time he violated it retribution was so swift and violent that it seemed as if his father's ghost had returned to slap him from the saddle.

Calves bawled, cowboys ran. The irons protruded from the fire like a spread sheaf of steel arrows. Men traded them in quick succession, switching a cool iron for a hot one. The brand was typically obscure to the uninitiated—a hook that looked slightly like a backward *J* hanging from a curve. The interpretation was Rocker Six.

In the corral the branding was a dance of weaving figures. In the center, with the milling herd, a cowboy in the position of honor rode his best horse to rope, the job on which all the rest depended. He was an athlete, grace in the saddle. The stiff loop flicked out like a snake's tongue, lashed around a calf's hind legs. The cowboy dallied swiftly around his saddle horn and pulled up in his own dust. The calf fell, bound, and the cowboy dragged him, a kicking sack, to one of three calf-wrassling positions on the ground, then released him, and the horse leapt back to the job. The wrassler heaved and the calf thumped on its side, legs stiff but flailing, like an animated table. The iron hissed, the smoke exploded from hair and flesh. Another piece of ear was sliced off. Another pair of testicles plopped in the bucket. In some parts of the West this harvest was dipped in batter and deep fried for breakfast the next day: Rocky Mountain Oysters. "How many calves we done?" asked a young wrassler, a high school sophomore, so eager he was naive. "A few," their owner assured, spitting in the manure.

Came a time the roper couldn't connect. He built a loop, threw, missed; built, threw, missed. All three pairs of wrasslers were out of work, calves finished, standing, stretching. The cowboy missed again, brought his horse up short, angry. You get asked to rope calves for somebody's branding, you get a compliment; you don't like to miss under critical eyes.

All the irons were waiting in the fire. The brander added wood. The man with the syringe examined his needles. Wally McRae with his spoon and his jar of salt paused with relief. Donald McRae methodically wiped his blade clean on a gray rail of the fence.

No one watched but everyone saw. The cowboy missed again. The horse whirled, foam glittered in the air, the cowboy threw an ill-formed loop and the calf scampered away, suddenly agile as a cat.

Josh McCuistion, owner of the Rocker Six, gave the roper a shout. McCuistion, old and lean and tough, wore laced shoes for boots and old tweed suit pants, used an old army blanket and a burlap sack for a saddle blanket, and carried oral tradition around with him like a pinch of Skoal behind the lip.

"Hey," McCuistion yelled. "Why don't you fight your horse? That helps. Leastwise, I seen a lot of fellas do it." It was an old hint, come up the trail from Texas.

The cowboy laughed, relaxed. Fighting his horse? Yeah, sure, that's what he had been doing. He turned, steadied, and threw a big pure loop and dragged the calf to the wrasslers. Iron! Knife! The smoke leaped out again.

At the end of the day a spectacular dusk in red and yellow approached the valley, laying itself down in long planks of shadow from the ridges. The cowboys quit for the day, laughing and tired; the fire glowed down to a mound of ash, the irons cooled, and the calves grew less noisy, accustomed to the new pain. Wally McRae loved it—the smell, the rich light, the weariness, the continuity that he did not yet fully understand but felt as strongly as an arm around his shoulders, assuring him that this had been going on all his life and would continue long after he was dead. He was taut with the joy of being a part. Eager to complete his contribution, he offered to pour water on the fire, but an older boy was sent to do it. Then he suggested he could gather up the irons, but that, too, was trusted to another. Finally, watching with the frustration of extreme youth, he saw something he could do, not only to spend his energy, but to show off his knowledge. The pile of ears remained uncounted in a corner of the corral. McCuistion, a friend of the McRaes since he came to the country, was busy elsewhere. At least, Wally thought, I can count the ears.

He squatted at the pile to begin. But eight deep he was interrupted by the nudge of a boot. His father, the short, solid man with wrists so thick he couldn't snap a western shirt's cuffs, had appeared from across the corral.

"That's none of your business."

The boy was startled. "But—"

"Those are Josh's ears."

"Then I'll help him."

Donald McRae shook his head. "No." The tone, Wally knew, was final. This was a rule of some kind, apparently; something absolute in his father's life. The boy was intent to absorb it.

"No," Donald McRae said. "It's none of our business how many calves he's branded this day. We don't need to know."

Not long before the branding there had been a spring storm. Lightning had flickered over the Sarpy Mountains and marching walls of rain had swept across the valley. A little tributary creek to the Rosebud had grown thick and rampant with the water. It had kicked stones and brush down its narrow bed, elbowing its way out into pastureland and nudging at the bases of the hills. At one bend it had so undermined a cliff that the old bank slid in and washed downstream as silt, leaving behind it a fresh new face of strata. And there, suddenly, like the *deus ex machina* revealed by the collapse of a protective flat of scenery, appeared the secret of this land, at which the burned scoria only hinted. Under the suddenly well-defined levels of sandstone and shale, tucked beneath the bank in a pitch-black slice so purely dark it looked like a profound gash in the earth, was the edge of a seam of coal.

If Wally McRae saw the coal beneath the bank as he rode home in his father's truck in the late spring light, he didn't pay it much attention. He knew about the coal. Northern Pacific was, after all, mining the stuff up at its town of Colstrip, where he had started school last year, and his mother's father had mined one seam of it during the depression; the seam was even named for him, the McKay seam. So McRae was practically related to the coal. No doubt if he saw the coal peeking out at him from under its brow of rock, he soon forgot the glimpse, as quickly as the little dry slides on the newly opened hillside smudged the lines of strata, dusted the black slice over and hid it from his inquisitive eyes.

6

The eleven-year-old girl was making mud pies in the reservation dirt when the summons came. And the change that was demanded of her was typical of her fractured life: One moment she was Marie Brady, a small, tan-skinned, Catholic girl, approaching first communion, playing happily, hands and arms streaked with mud; the next she was Wandering Woman, a warrior sister, symbol of the beauty and strength of the Northern Cheyenne tribe.

It was Busby, Montana, in the summer of 1950. It was the time of the Sun Dance. It had been seventy-two years since 297 Cheyennes walked out of the Oklahoma reservation where they had been sent after the Custer fight. Corralled there with a group of Southern Cheyennes, they had lost the will and the ability to live: Heat and malaria had been killing them, so they simply left, following their longing back north toward what one leader called "the land of pines and clear, cold rivers." They eluded twelve thousand troops, crossed two major railroad corridors, and nearly got through Nebraska before the group was captured and locked up at Fort Robinson. At the fort their stubbornness to return north was put to the ultimate test: When they refused to go back to Oklahoma they were locked up without food and water. After five days they broke out and all but a few were shot. The survivors continued north, and after the scandal of the Fort Robinson massacre became known, the Northern Cheyennes were given the current reservation here, 447,000 acres of forests, prairie and sandstone and scoria hills.

In the years since, while other tribes were disintegrating under the pressure of white American civilization, the Northern Cheyennes maintained their tribal integrity better than most, cradling it in their common pride in the long walk home, and in their love for the land that they had won. The language, at once sibilant and guttural, survived attempts at assimilation and so did the land: Unlike most other tribes, the Northern Cheyennes sold very few of those acres for which they paid so much. They considered it a cherished place. Years later they quoted a report of the Montana Environmental Quality Council to describe it:

"This subregion of a golden land," the report reads, "a land of buff-colored sandstone cliffs, ochre-tinted Ponderosa pine bark, and expanses of yellow grass . . . is intimate with a feeling of closeness between man and earth."

"Up north," said a Northern Cheyenne woman as she lay dying in Oklahoma, "the pines make a rustling sound in the wind, and the trees smell good." Over the years, while they waited in poverty and isolation and sometimes despair for the next test of their affection for their homeland, the Cheyennes did not forget that fundamental longing for the hills and the trees and the clean air of home.

And now it was time for the Sun Dance, the annual four-day celebration of life and God—Maheo—and the people had come from all over the reservation by horse and by pickup truck to be made new by the sun. The ceremony still held its ancient power: If Sweet Medicine's predictions were true, if the time was still coming when the people with hair all over their faces and white skin would control the Cheyennes, when the members of the tribe would marry their own relatives, become crazy, and forget their history, it had not yet fully arrived. Is'siwun, the Sacred Buffalo Hat, still hung in her buffalo hide sack in the Sacred Hat Tipi under the vigilant care of the keeper, and the Sun Dance still renewed hope every year.

The town of Busby was hot and dusty, a little village of gray log homes and multicolored government housing, spread on a high grass plateau like a handful of Indian corn poured on a plate of light. On this day in 1950 it was bustling and noisy with Indians from all over the reservation, the people from neighboring towns living in tents in the field near the tipis, where the Sun Dance Medicine Lodge would be built on the second day.

Here the dancers would face the rising sun on the third day, blowing their tufted eagle-feather whistles to greet the morning. Here the warriors would count coup for old battles, only now they would tell stories of fighting Germans or Japanese instead of U.S. cavalry. "I was at the Battle of the Bulge, drove a light tank, got a purple heart." From here the Sacred Woman and the Instructor would walk, to lie together on a buffalo robe in the field and renew the life of the people with their union. And from here the dancers would run on the fourth day exalted by sacrifice and thirst and, perhaps, self-inflicted pain, dash to the four points of the compass and then return to the center pole of life, to collapse in exhaustion and joy.

But in Marie Brady there was confusion.

"A Catholic sins against faith by not believing what God has revealed, and by taking part in non-Catholic worship." Her little blue catechism, with the underlined words and questions and answers, was not here; her grandmother had probably arranged to leave it behind, and she could not remember the precise words or whether they applied here or not. She had been baptized a Catholic when she was four, and she went to school at the St. Labre Indian Mission School at Ashland, on the Tongue River. So the rites of the church had surrounded her almost as long as these. What was real? She kept her conflict to herself.

At school the Cheyenne language was exterminated like an evil weed poking up through concrete. You were made to kneel in penance for speaking it, or you were slapped. "How many times do I have to tell you, Marie? We do not speak Indian in this school." But today the language was all around her, like a flood of disobedience or daring or ignorance or honesty. And now she was being taken from her harmless nonpartisan mud work on the edge of this worship of hats and arrows and other things that the teachers might not consider holy, and was suddenly in the middle of it.

Her grandmother's sister-in-law had taken her, grabbing her by the arm and hauling her to the family's canvas tent. With rough efficiency the woman mopped off the mud, ignoring Marie's protests, and braided her hair into two tight plaits, jerking her head with each braid. Then she dressed her in a heavy buckskin dress, stiff and beaded.

There were examples in the catechism of little girls who properly offered God the supreme worship that was due him: "The girls tried to get Martha to go to a fortune-teller, but she would not go. Julia's father is not a Catholic. Last Sunday he asked her to stay home from Mass and go for a walk with him. Julia went to Holy Mass first." Marie stood still as the dress was arranged, the beads adjusted. There was a role coming for her in all this. She kept her nervousness to herself.

They were outside the tent again. Suddenly Marie's great uncle was there on his horse. He leaped off, a large but graceful man, removed the saddle and replaced it with a wool blanket and a shawl. Then he lifted Marie on, tying another shawl around her. The girl was unresponsive and almost surly in answer to their mood of urgency. What was all this for? What was going on? Why her? They did not explain.

Then she heard the voice of the crier, the man who walks through the village and shouts matters of importance to the people. The voice was clear: It penetrated the ambient chatter. It was calling her name. Wandering Woman. Wandering Woman!

She looked up. From high on the horse's back the whole field was suddenly visible, a bright collection of tents and tipis, horses, men and women, running children. It was a great mass of people like her, a gathering with one intention, to celebrate their particular kind of life. At school the white people called the Cheyennes the Race of Sorrows; here there was a jumbled noise of joy. What was real? The crier called her name again, and it pierced through her reluctance like an arrow of glory. The horse moved slowly through the crowd like a ship. She was above all, watching. She kept her wonder to herself.

They came to a tipi standing apart on the grass. There were people waiting for her; she was ushered inside. Here several men shook her hand and smiled. They called her "sister." And slowly she understood. She was being initiated as one of four sisters to the Elk Society, one of the ancient Cheyenne warrior societies. This was the society whose responsibility it was to protect Is'siwun; this was the society to which her great-great-grandfather, Chief Little Wolf, belonged. Little Wolf, with Dull Knife, had led the Cheyenne people out of captivity in the sweltering prison of Oklahoma back north. There was honor here in this tipi.

The crier was out in the camp. His voice was distant, the names indistinct. Then the people he called came running. Marie's uncle awaited them, smiling. Now he gave away the wood and the shawl— and the *horse*. All to honor her, the girl wrenched away from mud pies.

The Cheyenne language flowed around her, her own language, the sin, the safety of it. What was real? She sat in the position of honor, the men of the Elk Society all around her, great shapes in the hot muted light of the tipi. Shadowed, heavy faces. Men wearing multicolored shirts and Levi's. It smelled of man sweat and cedar smoke. These men were all now her brothers; according to the custom, she might now never marry a member of the Elk Society. She was a symbol of all that was good among Cheyenne women. It was a position of great and somber meaning. She kept her fears to herself.

The heat of the cedar smoke bit her lungs, but she did not wince or cough. She could handle it. She knew how to handle them all, the symbols; those she had and those she had not tasted. "Do this in remembrance of Me." Blood wine, wafer body, smoke—breath? What was true?

"Are all obliged to belong to the Catholic Church in order to be saved?" the catechism asked rhetorically, and answered: "All are obliged to belong to the Catholic Church, in some way, in order to be saved." Perhaps it is the greatest fear of all, the childhood fear of being lost forever. "You can tell by the way Grace kneels that she believes Holy Mass is the same sacrifice as that of the cross." Five years before, Fred Last Bull, priest and future keeper of the Sacred Arrows, had had his chest muscles pierced with skewers in the old custom, then tied the inserted skewers to the center pole of the Sun Dance Lodge and danced, stepping backward until the pins pulled through the skin. Marie remembered the story vaguely, the blood, the sacrifice. What was real? She kept her loneliness to herself.

In 1957, in Busby, Montana, Fred Last Bull interpreted the prophecy of Sweet Medicine, in an interview that would become the preface to a book written ten years later, *Cheyenne Memories*.

"He said the white men would be so powerful, so strong," Last Bull said, "they could take thunder, that electricity from the sky, and light their houses. Maybe they would even be able to reach up and take the moon, or stars maybe, one or two. Maybe they still can't do that. . . .

"They will be powerful people, strong, tough. They will fly up in the air, into the sky, they will dig under the earth, they will drain the earth and kill it. All over the earth they will kill the trees and the grass, they will put their own grass and their own hay, but the earth will be dead—all the old trees and grass and animals. They are coming closer all the time. Back there, New York, those places, the earth is already dead. Here we are lucky. It's nice here. It's pretty. We have this good air. This prairie hay still grows. But they are coming all the time, turn the land over and kill it, more and more babies being born, more and more people coming. That's what He said. . . . It's all coming true, what He said."

7

In 1958 Stewart Earnest White, like the city of Butte, Montana, in which he lived and worked, was growing more grim as he grew older. The many-layered city of dark brick buildings and yellow smoke was fading from its days of ebullience when the hill on which it stands was called the richest on earth. The city was still dominated by the tall wooden sentries called headframes or gallows, whose whirling steel wheels told of travel somewhere beneath the surface, deeper, miners were proud to say, than a mile down. But the Anaconda Company, winner of the Great War of the Copper Kings in the early twentieth century, had begun, in 1955, an open pit mine on the side of the hill, and the pit was growing. In time, long after Stewart White was gone, it would gnaw into the heart of the city, unfolding like a huge ear in the earth in which, even on a Sunday, you could hear the sound of engines. In 1958 the Berkeley pit was small and the town was hardly dying, but the great days when, it was said, a Cornishman could be put aboard a ship at Liverpool with a tag on his coat labeled "To the Seven Stacks of the Neversweat" and be delivered without question to that well-known mine in Butte, were gone.

Every Friday the Anaconda Company foreman tacked a small list on the bulletin board that stood outside the steaming, sweaty room known as The Dry, in which the men coming up from the day underground showered and changed. This paper ranked the men the way sportswriters rank college football teams, except that position was based on tons of ore moved rather than touchdowns. There were, no doubt, men in every shift who hated the rating system as a needless vanity; everyone was a contract miner, paid by the ton instead of the hour. Since the paycheck reflected their effort the ranking was meaningless, just another gimmick with which Anaconda, the company which ran the state of Montana, fed artificial incentive into the veins of its underpaid troops. But White could not resist the competition, and the best Fridays were the ones in which he could go home to his tiny frame house, where his son slept in a room not much bigger than a closet and his daughter slept on the couch, and tell the family: "This week we led the board."

But there were also days when White returned to the small home that he wasn't sure he could pay for, watched his wife work without complaint in a cramped kitchen, and finally muttered: "Ah, I'm nothing but a dirty miner."

He had tried so many other jobs. He had sold life insurance, then went back to mining. He tried to ranch on a little farm not far from Butte where the family lived without plumbing or electricity in a two-room log cabin and where he had to work as a part-time section hand on the railroad. But nothing ever seemed to suit him as much as mining, mucking out the copper or the manganese ore for the Anaconda Company in Butte.

"I think he felt comfortable there," his daughter, Sylvia, remembered years later. "And then when he'd say he was nothing but a miner it would make Martin and me so mad because we felt that it didn't matter. But to him it really did; he expected a great deal of himself, and he just never made it."

In 1958 Stewart White was forty-six years old and had gone back to mining for good. There would be no more attempts to break away from this work which so gripped him and yet so fed his despair. Perhaps he was feeling old, faintly aware that he did not have much time left, or perhaps he was just tired; but the optimism that was once his dominant mood was now deeply eroded by bitterness at the company's treatment of its men; by what he perhaps considered was his own failure; and by his deep and extensive reading, which had begun to persuade him that there was little promise that humankind could ever escape the endless round of waste and cruelty and greed that seemed to dominate its history.

But as White became more discouraged, his children were growing up cheerful and confident, almost as if he had robbed himself of hope in order to give it to them. They were intense and talented kids; speedskaters, it would turn out, of Olympic quality. The competitiveness that made Stewart White struggle to muck out more ore than any other man was in their blood, too.

White liked either three thousand feet of stone over his head or nothing at all, and when he had time off he took his son, daughter, and wife out into the broad humps of the Rockies around Butte, to places where the demand for mine timbers had not yet destroyed the forests. In summer the family picnicked in the Tobacco Root Mountains, driving in its 1950 Plymouth up Brownback Canyon where the South Boulder River runs swiftly through the trees. And in the fall he and Martin would go hunting. On those short, intense trips into the wilderness, White and his son liked to argue about all the conflicts of the world.

Early in November 1958, White and his son took such a trip, driving north out of Butte into the heavily timbered country of Thunderbolt Mountain. It was familiar land; good elk country. They had hunted there before. And the weather was perfect: a driving blizzard.

"It was snowing and blowing and the Good Lord was really getting with it," Martin remembered. "We always liked to hunt on the worst days, because the snow kills your scent and the wind makes so much noise they can't hear you coming."

But when the old Plymouth reached the first of two canyons the Whites liked to hunt, a little of the grimness that usually left Stewart White on his trips into the country returned to his face. The draw was already occupied; a jeep was working its way up the slope where they would have had to walk. They could dimly see it, a gray, groaning shape in the falling snow. White's frown deepened. It was as if he found the pressure on his life by too many people intolerably heavy, the way another man might instinctively feel the weight of the earth deep in the mine.

"Goddamn four-wheel drives," he muttered, and drove on.

The next canyon was empty, and Martin soon forgot the intruding jeep. Hunting was as much his passion as his father's, and he took the stalk seriously. He hunted with an old Winchester 30–30. The weapon, he often thought, had a will of its own, a kind of absent-minded inaccuracy. He blamed its inconsistency on the bore of the barrel; his father called it buck fever.

They hunted without telescopic sights. They couldn't afford them, and Stewart White, typically blunt, didn't like them. "Open sights are all you need, kid, faster in the timber."

The hunt led them about four miles up a slope of canyon and over a ridge, while all the time the sound of the jeep grew and faded in the background, blown by the wind. But when they came down over the ridge it disappeared. And there, in a clearing, distinct even in the blizzard because he was so close, was the biggest bull elk Martin had ever seen.

Stewart White wasn't a trophy hunter, much to Martin's disgust. Old meat is tough meat, and the family had to eat what it killed. Once, Martin remembered with a pang, he and his father had climbed over a saddle and looked down at an enormous buck mule deer. Martin's eyes got very large, and he had aimed the Winchester, but his father had pulled the barrel down and spooked the deer. "No, no, no," he had said. "You can't eat *him.*"

But this time there was no restraint. The elk had not noticed them, although he was barely fifty yards away. He was standing in the clearing beside a lone fir tree, nibbling on the side of the tree, head turned. The image would remain in Martin's mind for years. Martin froze where he stood, at the edge of the clearing, and he turned to whisper to his father.

"There's one!"

He never forgot his father's reply, although at the time he didn't understand it. The elk stood as big as a cliff, a great bulk of fur and antlers, brown against the gray and green of the forest.

"I don't see him, kid," Stewart White whispered back. "You'd better shoot."

Martin put the 30–30 to his shoulder, the stock cold against his cheek, snowflakes nipping at his eyes. He aimed for the heart, and fired.

As the shot echoed briefly out into the blizzard, the bull elk fell, but he was not dead. The 30–30, or Martin's excitement, had robbed him of a clean kill. The elk rolled downhill, away from them, hit the edge of the woods, a mass of young trees tangled in crisscrossed deadfall, bounced to his feet again and disappeared into the forest. As Martin and his father began to run down the hill, following the trail of blood, they heard the sound of the elk, crashing through the brush, the rush of the wind and of their own breath, and, in the distance, the thin roar of a jeep.

"I'll never forget him saying that," Martin White said much later. "I thought even afterwards, even then I thought, why, Jesus, he can't see it, the goddamn thing takes up half the park. But that was Dad.

He'd much rather see me get it than himself. He was—he really would rather have his family get something than himself. He never had much in his life."

Stewart White loved mining but he distrusted The Company, as Anaconda was known in Montana. He was a free man, "so straightforward and so honest," Martin said, "that sometimes I used to just cringe." But The Company controlled him, and he saw no heart in it.

"It was my dad's feeling about corporations that people had to be protected from them because their objective really was contrary to looking after the people," Martin remembered. "Their objective, he said, was making the most possible money for themselves, and to do that they had to keep the cost of having an employee on their payroll as low as they possibly could. And he had a deep distrust of the Anaconda Company.

"I was a senior in high school and I knew just everything there was to know in the world. I thought that both industry and government to a certain extent, were good. And my father felt that there had to be some protection of the people against both industry and government. I didn't—I wasn't farsighted enough to recognize how difficult it is for a company to be concerned about anything but making money.

"And so we'd end up in these discussions. And that, I guess, is the thing that I always used to love about Dad, was because he and I would have these discussions, just really get with it and afterwards, why he'd grin; he'd think it was kind of fun to argue with his kid."

In the forest near Thunderbolt Mountain, Martin White and his father sat on a log beside the dead elk. The open cavity where the entrails had been steamed up into the snowy afternoon. The two men's hands were still warm from the work. It had been all they could manage to roll the elk over to gut it, and now they were faced with the problem of packing the meat out, up and over the hill and four miles back to the car. The elk's final sprint, before the animal weakened enough for them to hit it again, had taken it deep into a tangled wilderness of downed trees in which even an unburdened man would have trouble walking.

"Damn, kid," Stewart White said in mock anger to his son, who was still eyeing the enormous antlers, "when are you gonna learn how to shoot?"

For a while they were silent, absorbing the moment. The snow and the wind and the huge brown animal, stinking of clean death. Nothing could be more primitive; nothing could be more natural, for a man and his son, gathering their winter's meat. Tomorrow they would return, with friends and packboards, and carry it out. Now they could rest.

But in the distance, above the wind, they could hear again the sound of the jeep, gnawing minutely at some snowbank in the other canyon. Stewart White stiffened almost invisibly. The pressure returned.

"Damn jeeps," he said finally. "It's bothering me."

Martin, still tingling with adrenaline from his triumph, came right back.

"That's ridiculous," he said, his tone bantering. "What the hell's it bothering you for? Just some guy gettin' out, they aren't hurting anything."

But Stewart White was serious. "It's bothering me," he said again. "Animals don't have a chance anymore with all the four-wheel drives around."

"Probably some old fellow who couldn't get an elk without a jeep."

"Well, hell, if he has to tear up half the countryside to get an elk he shouldn't get one. A law should be written to stop vehicles from leaving developed roads."

"You don't want more government, do you?"

"Hell, no." White considered it a moment. "No, I don't. I don't want any more government. I just want the government we have to protect what little open country's left in America. The way we're going, ten years from now there won't be any part of this country you couldn't drive to. All the game will be gone, the timber will be gone, and there will be roads and erosion everywhere."

Now Martin got serious. The pressure on his father seemed to be selective, like vertigo. It didn't bother the son. He saw nothing sinister in the driver of the jeep; he saw a friend from school or someone's father where Stewart White saw a threat.

"Don't you think it a bit selfish to want all this country for yourself?" he asked him.

"Selfish?" White paused. The jeep muttered again. "Yeah. Maybe. Somebody has to be selfish about this land or there won't be any left to be selfish about."

"I don't think we'll let it get that far. We'll protect it—"

Stewart White laughed shortly. "Man is basically greedy," he said. "His greed will drive him to cut every last tree, build roads up every valley, and cover the land with cities. Look at the Roman Empire. Destroyed by greed. What drove Spain and England to America? Dreams of riches: greed. John D. Rockefeller and others like him run the country because of their greed."

The dead elk steamed beside them like a meal ignored. They went on.

"I will have to concede," Martin said, stuffy with words in his youth, "that greed has got something to do with civilization. But don't you think there's some peripheral good that comes from this advancement? All the good people that work for the good of their fellow man?"

"Good people," White let some of the pressure out, a little spurt of anger. "Good people. There's *too many* people in this world. And it's going to get worse. And in your day—in your day it's going to be the greatest single problem you'll have to face."

But Martin White laughed. He didn't believe it. And slowly the sound of the jeep, that so terribly nagged his father about the growing pressure of human beings on the world, faded away into the soft storm.

8

The Northern Pacific North Coast Limited rambled through the late afternoon on the Great Plains, rolling east behind the two new diesel locomotives that had just replaced the last steam engines on the line. With a hoarse, self-confident roar of internal combustion the train thundered along the Yellowstone, through the quiet western towns of Pompey's Pillar and Hysham, shaking the earth with its power and leaving behind the sharp smell of burned oil instead of the soft sulfur smell of coal. Approaching Forsyth the train barely slowed at the switches at which the tracks led up Armell's Creek to Colstrip, twenty-nine miles south. Those tracks, on which once came the coal to power trains into all the corners of the Northern Pacific's little empire, were already rusting as the diesels passed by, arrogant with their more-convenient fuel.

At the end of that line the huge dragline that Fred W. DeGuire once described with such enthusiasm stood in silence, unmanned; and a million and a half tons of subbituminous coal, the earth already stripped away from it, lay weathering under early snow. In the history of the power plant that would later rise near these banks of abandoned coal, it was a hiatus. The plant, unborn except in DeGuire's fervent imagination, could have expired here as only a dream. DeGuire had gone long before; curiously, this man who had done so much to promote the growing dependence of human beings on technology recoiled from it in the end: When the heart attack that killed him began he refused to enter the doors of a hospital and so died that night at home. The idea of the power plant could have vanished with him.

But riding on the North Coast Limited that day, perhaps reading *The Case of the Lucky Loser*, a novel by his favorite author, Erle Stanley Gardner, while the train took him to St. Paul, was the man who, more than anyone, would help bring DeGuire's prediction, and the power plant, back to life.

For the small but audacious utility that called itself the Montana Power Company, of which this man, W. W. "Dutch" German was a vice-president, this was a year of change. For almost fifty years the company had relied on one of the easiest ways of making electric power—falling water—to supply its customers across the state with the energy they needed, and now future supplies of this commodity seemed limited. Two major new dam and power plant sites were being considered in northern Montana, but they were stalled because the federal government wanted to build a larger dam on the river near them. Several dam sites were also being studied in Hell's Canyon on the Snake River in Idaho, but there again conflict between federal and private projects held them up.

But the use of electricity was continuing to grow. This magic, sparkling substance that could be piped hundreds of miles and could run anything was flowing into homes and industries just about as fast as Montana Power could pump it out. And, like any good capitalist who was marketing the gadget of the century, Montana Power wanted to make and sell more. In 1958 the average household served by Montana Power used 3,736 kilowatt-hours a year, and the company wanted to see the level of use increase. So it had developed an advertising campaign called "Electrical Tuesday" in which it ran ads in newspapers once a week describing the many new uses to which electricity could be put. With occasional promotions like these, Montana Power hoped that the average household use of electricity would double in twenty years. The only trouble was finding the right source of energy with which to make the electricity.

Just seven years before the world had been delighted to hear that the Atomic Energy Commission had succeeded in providing the electrical needs of a tiny town in southeastern Idaho with power from a nuclear reactor for half an hour, and the potential of atomic power was now spoken of with the same kinds of neon adjectives that DeGuire once used to describe Colstrip. It was hardly imagined that the bold hopes for an atomic answer to all energy problems would ever fall into a similar sort of decay. To keep up with the times, Montana Power was participating with fifty-two other utilities in a 40,000-kilowatt nuclear plant at Peach Bottom, Pennsylvania, but the company's officers did not plan any serious venture into that still-mystical field for some time. So when the company could build no more dams it would have to burn something to get its power: natural gas, oil, or coal. And when Northern Pacific got tired of shoveling coal when everybody else was pumping diesel and put its Colstrip mining operation up for auction, Montana Power got interested. Its interest grew when nobody bid.

"That's when we had the chance to go in and really make an arrangement," a Montana Power officer said later. "When it turned out that we were the only ones that were interested in it, that made the bargaining much easier."

So Dutch German, a buoyant, slender man characterized by his grin and his habit of stating a proposition in two parts, never in one or three, began to travel back and forth from Butte to the Northern Pacific's St. Paul headquarters, negotiating for the coal mine that nobody else wanted.

What he offered and what Northern Pacific wanted are buried in the official tar pits of reticence in which old corporate records tend to rot. The bones that can be excavated are bare results: After a half year or more of bargaining, German persuaded Northern Pacific to settle for a lease-purchase arrangement in which Montana Power became the owner of the town and equipment at Colstrip and of a thirty-year lease on a total of about sixty million tons of coal. The total cost of the acquisition was not disclosed.

No one was aware of the cost of the negotiations to Dutch German himself. Without telling even his wife, German extended the first trip to St. Paul and continued on to Rochester, Minnesota, to the Mayo Clinic, where he was diagnosed for the cough that had bothered him for months. It was lung cancer. He was operated on immediately, then returned to work, shuttling back and forth to St. Paul by aircraft and train. Shortly after Montana Power closed the deal with Northern Pacific, German extended another trip and was told at the clinic that there was no additional evidence of cancer. But the celebration, if he indulged in one, was premature. By spring, no longer able to hide his illness, he quit work and entered a hospital in Billings, where he died on April 1, 1960.

There was no particular sense of irony in the sorrow Dutch German's employers and friends at Montana Power felt at his death, no bitterness that the company gave its future to the care of the one man who knew he would never see it. The purchase that he engineered, of a dormant coal mine and a yard full of huge machines pickled in silica gel did not, in the spring of 1960, seem that important.

9

In the summer of 1962 the town of Colstrip was quietly fading away. It had never been a city; even in the hard-driving days of the 1930s and '40s it had just been a few square blocks of identical company homes and a cluster of offices that looked like barracks. Now, with the Northern Pacific mine shut down, the numbers of boarded windows exceeded the open ones. In the mines around the town the coal that had been laid bare but not yet mined rotted in place like old black bone, corrupted by wind and rain. Colstrip had become a cow town, with its grain elevator and stockyard. The cattlemen met in the schoolrooms evenings to talk about roads and import tariffs, and their wives met in the little combination store and soda fountain in the school's basement. About the only miners left in town were two men named Nolan Fandrich and Obert Rye, who were taking care of the town and the machinery until somebody might have use for it again. The two men told the occasional casual visitor rich and hopeful stories of the great past days of coal and power.

One day in 1962, however, Colstrip had a visitor who was not interested in nostalgia. The weather on that day was hot and dry— typical late summer. The fields beside the cracked road that leads down Armell's Creek to Forsyth were already a burnished brown, and the dirt roads lay smothered in dust. In early morning and late evening the plumes thrown up by pickups on the ranches caught the low sun and blazed like ridges of flame, then gently dissipated to fog the landscape in amber light.

In mid-afternoon a large Ford automobile, rented from the agency in Billings, pulled slowly out of Colstrip onto the old road and headed north, heaving on sluggish shocks to the contours of the pavement. Just north of town it slowed as if on impulse and pulled in to the

39

dirt parking lot of a tiny building called the B & R Bar. Two men got out, dressed in suits and ties, laughing together, and went into the bar.

Inside it was dark, but hardly cool. Not even flies stirred. Even in the mirror back of the bar the light seemed hazy, thick with the heat, the room obscure. As the eyes of the two men became accustomed to the darkness, they saw that the room was empty, pool cues racked on the walls. But shortly after they entered, ringing a small bell, a man in shirtsleeves appeared from a side door and, upon request, produced two cold glasses and a couple of bottles of beer.

The bartender was reluctant to leave, with two strangers in city clothes to talk to, so a conversation began. Weather too hot, not enough rain; nothing like the drought of '61, though. The words seemed too loud in the empty room; the men instinctively talked quietly, as if the silence listened. Business? Slow as hell. Ranchers, cowboys around here work late all week then go into Forsyth on Saturday night. Nothin' like it used to be when the mine was goin'.

This seemed to interest the larger of the two men; the rounder one who looked like a young and friendly grandfather except for slightly hooded eyes. "Really," he said. "Used to mine coal out here, then? Well, what's happening here now?"

The bartender's mood turned slightly cynical.

"Guess Northern Pacific sold the works to Montana Power Company. Can't have got much for it or the power company got took."

"Um," said the large man into his beer. "What do you think they'll do with it?"

The bartender snorted.

"Them? Nothin'. Nobody's gonna do anything. What *can* they do with it? Put the coal on the roads in the winter?"

"Well," said the big man, idly swirling the remains of his beer. "Maybe they're going to build some steam plants and they'll use the coal for that."

"Naw," the bartender said, wiping up a nonexistent stain at the near corner of the bar. "They got all the water power they need. Hell, if that happens it wouldn't be in our lifetime."

The two men finished, paid, and left the room; the silence closed in behind them like a swinging door. In the car the younger man smiled at the older one.

"How come you didn't tell him who you are?"

The other man, George O'Connor, a vice-president of Montana

Power Company, down for a visit to the Colstrip property, smiled, the eyes half-closed. " 'Cause he wouldn't have told me what he did."

Early in 1963 Montana Power ordered a boiler capable of providing enough steam to produce 180,000 kilowatts of electricity. The boiler was to be designed to burn coal from Rosebud County, Montana, and was to be delivered to Billings no later than the fall of 1967.

10

Studies now underway by Montana Power Company and four other Rocky Mountain utilities indicate that steam-electric generation utilizing Colstrip coal could become increasingly important within the next 10 years.

—*Forsyth Independent*
August 30, 1962

On the Rocker Six ranch down on the Rosebud it was starting to get dark on a featureless overcast December evening. The sky, crumpled gray with clouds, matched the land, rolling gray with stubble and snow and sagebrush and ridges dark with blue-black trees. Through the pasture near the house and barns, a big sorrel gelding loped, his rider holding slack reins. The horse kicked up tufts of crusted snow—pale, glittering splashes in the dusk. The rider was at ease, slender and erect, with a dark blue scarf wound tightly around his neck and a lean, moustached face almost indistinguishable in the heavier darkness under his hat. There was just a glint, a hard flicker of eyes. The animal gathered speed now, eager, but the rider made no move to hold it back. For the first time in his life, Wally McRae was letting his horse run home.

It was three days before the end of 1963. McRae's father had been dead for three and a half years. He had left a lot of empty jobs behind: secretary-treasurer of the Greenleaf Land and Livestock Company, first vice-president of the Montana Stockgrowers Association, director of Forsyth State Bank, director of Miles City Salesyard, director of Tongue River Electric Co-op, and member of the Rosebud County Fair board. The kind of jobs that line the walls with more certificates of appreciation than dollar bills.

Now Wally was home with an honorable discharge from the navy as a lieutenant j.g., a Pennsylvania-born-and-raised wife, Ruth, a son, Clint, another child on the way, and a huge new ranch. Huge, at least, to him. In these parts thirty thousand acres is just a good working outfit, no empire.

The horse ran. McRae was thinking, letting the wind go by, of other things than old warnings. Contemplation was always his vice. He had owned the ranch for two weeks and, smart, well educated (B.S. in zoology from Montana State University), and born to the role as he was, he was nevertheless overwhelmed with the size of the land and the debt that accompanied it. In November of each year, for the next thirty years, he must pay the bank twenty-one thousand dollars. And that was just the land; the seven hundred head of cattle and the ranch equipment were all extra. Still, in spite of this new burden, a heavy one for a man of twenty-seven, there must have been some elation hidden under that hat as he rode home on his land, or else he wouldn't be letting the horse run.

This was the end of a long search. When his father died suddenly at fifty-eight, McRae sold his interest in the Greenleaf Land and Livestock Company and went out to look for his own outfit. The roads are long in the West between towns and ranches and he put a lot of miles on his '61 two-door Ford, just looking. There was a place near Cody, Wyoming, with a waterwheel for electric power. There was a place near Big Timber, in the rainshadow of the mountains, but tall larkspur grew in profusion in the coulees. Green death for cows. And then there was that outfit near Dixon.

They found it at the end of a long dirt road, one of those twisting cat tracks pasted insecurely to the side of a hill above a river. The house was dirty, a peach-colored stucco, and the yard was full of uncleaned bones and overturned cars from a previous decade. Ruth got out of car and recoiled, but since she is a person of considerable patience and restraint she did not demand to leave immediately. The whole place smelled like an abandoned root cellar into which something large had crawled to die. You had to duck to go into the house, if you dared.

The only good thing about the outfit was a well-developed irrigation system using water from the adjacent river, which promised to provide a good hay base, and a large raspberry garden.

Since they had driven three hundred miles to get there, the

McRaes explored carefully, noting the excellent bottomland hay ground and the scenic value of the river; anything to offset what Ruth later called the overwhelmingly dreary mood of the house and yard. Then when they were leaving the realtor sprung the final blow.

"Well, ah, you've heard about the dam?"

"No. Dam?"

"Well, ah, if the proposed dam up here goes in, they're going to flood everything on this place that isn't standing on end."

So the search had continued. McRae went to look at an outfit up near Opheim, near the Canadian border, but up in that country the wind blows like a long knife and the only place that even the snow can find shelter is in the lee of a cow. Nothing quite compared to the subirrigated meadows, the high grasslands, and the baked-red scoria cliffs of home. And every time McRae returned from these trips he'd compare notes with a neighbor, a widow who now owned the ranch that used to belong to Josh McCuistion, and who was trying to sell it. It seemed too expensive for McRae, so they'd just visit.

"Finding anything yet?"

"No. Got a buyer?"

"No. Had a fellow over yesterday; son of a gun had more curiosity than money."

But the more McRae thought about it the more he liked that ranch, though it was bigger than what he thought he could buy. So he went to the banks and the Production Credit Association and finally was able to come up with enough money to buy the Rocker Six.

The deal was complex. It involved a complicated division of oil rights and royalties on the property. But it was 1963, the town of Colstrip just ten miles up the road was nearly a ghost and, as McRae remembered much later, "during this whole, long, involved process, the word *coal* was never mentioned."

So now he was returning from his first ride alone on the outfit. It was December 28, 4:00 P.M. All around him the land stretched, his own: cliffs dark in twilight, pastures and hayfields laid in patches of lighter gray along the narrow forest of box elder and cottonwood that wound through the valley floor and beside the black, silky waters of the Rosebud. Cattle sprinkled the lower slopes, drifting along in groups like distant clumps of tumbleweed rolling before a slow wind. Nearby a straw stack of interlocked bales was an abrupt boxy shadow. The slight glow from his house, beyond a rise, was the only artificial light that penetrated the night.

Earlier, moving through the cattle, searching for one cow who was being overwhelmed with lice, and thinking about the size of his ranch, McRae had been insecure.

"My God, McRae," he remembers thinking, "what have you gotten yourself into? Are you going to be able to handle it?" The man who had been running the place would be leaving in a month, the calves had not yet been sold, winter had begun. But now there was a different mood. He might be in deep but he was in it alone; no longer just a cog in the family corporation that was Greenleaf Land and Livestock. He had taken a full share in that continuity of ranching life that stretched back to his grandfather and ahead farther than he could see. And he knew the land better than he knew himself. So he threw the slack to the horse and Bonine ran, the cool fresh wind in his face, up the hill toward home.

But there were two gates; one led into a hay meadow, the other to the horse pasture. Unaccustomed to the layout of his outfit, McRae headed for the wrong gate, then discovered his mistake. Without slowing Bonine, he turned him with the reins. But the footing here was slick and the horse went down. There was no stumble, just a sudden crash. The horse fell left.

McRae was unable to jerk his leg free. The horse slid on the rough ground; McRae's leg, pinned, rolled over and over beneath it. There was a burst of pain.

Bonine was up, breathing heavily, blowing spit. McRae lay sweating and shivering in the snow and mud of the Rocker Six.

11

Many years later, testifying before a congressional subcommittee, Wally McRae would say: "I don't know if we like adversity or what, but we *love* this country." But on this night on the Rosebud he had not yet reached the point at which he would have to define his relationship to the land. All he had to support him now was tradition.

"All of my life I can remember that the ranch was *the* most important thing," McRae said later. "It was more important than comfort or happiness or *anything*. It was more important than family. It was more important than marriage. It was more important than religion. It was *absolutely* the *only* important thing in the world."

The winters are seldom gentle in eastern Montana. The chinook wind, the soft and sudden breath from the south, is uncommon; more often the air is drawn straight from the Arctic, and the shelter provided by trees and coulees only offers deeper drifts. Cows cluster and steam their body heat away, and if you know one is weak you bring it in to shelter or it dies. When the wind blows pellets of hard snow on air chill enough to freeze the living flesh, some ranchers let the poor ones die. Donald McRae had not been among them.

"I can remember him going out and riding in the winter time to get in a thin cow," McRae said. "And he'd come in and his face was frozen. He'd go out knowing full well that he was going to come in and both sides of his cheeks would be absolutely frozen stiff and they'd be white and they'd peel and break and bleed and be raw. He'd know that that was going to happen to him, but that cow was that important to him that he'd go out and get her in the winter time."

It was 6:00 P.M. on the Rocker Six. The night had closed down hard on the valley of the Rosebud, although the black land made the starlit clouds slightly luminous. From where he lay on his back by the gate into the horse pasture, McRae could barely see the glow from the house a half mile away. Occasionally, too, he saw a small arrow of light moving along the county road beyond, and heard a distant rumble from someone's pickup running home.

McRae had found that by lying on his back in the snow and mud and hunching slowly along on his elbows he could make progress against the drag of his broken leg. And so he had reached the gate in an hour's effort. If he could just get it open he could let the horse through, and the empty saddle would be a message. But he could not raise himself high enough to unlatch it. As he pulled himself up on the wires the useless leg bent and twisted like a sock full of rotten oranges, and every movement blasted him again with pain. He slumped back down to the cold ground, shivering. He was not afraid of death, just of failure.

"I have really loused up," he remembers thinking. "Now what? Of all the dumb things to do. Your dad told you all your life never run a horse towards home. You do it one time in your life when you really can't be stove up and here you are laying out here. What a stupid thing to do!"

He lifted himself once more on the fence, eyes pinched shut against the pain, and felt for the loop of wire that held it shut. He had it, but he had no leverage. He heaved it upward, but the leg flopped askew and with the shot of agony he let himself back to the ground. Straining to hear, over the rattle of his shivering, the sound of his hired hand ending his nightly routine by closing the doors of the equipment shop, McRae let out a long, warbling shout, the kind of yell he used to call cattle to feed across the pastures. The sound leaped out and vanished without an echo. He tried it again and again. But each time, after the yell, the silence seemed more profound.

He saw the bulky shadow of the straw stack nearby. He had matches in his pocket, though his cigarettes were gone. If no one found him by ten o'clock, he decided, he would light the stack. He relaxed on the ground to wait, yelled again, and continued to yell at intervals. At last there was an answering shout.

A few days later, as McRae lay, thickly casted up to his hip, in the hospital bed in Forsyth, Ruth had a miscarriage. And as they spent New Year's Eve in separate rooms in the Rosebud Community Hospital and began their first year on the Rocker Six in enforced absentee ownership, McRae thought: "It just can't get a whole lot tougher than it is right now."

12

COLSTRIP LIGNITE DEPOSITS WILL
FIGURE IN MONTANA'S INDUSTRIAL
DEVELOPMENT PLANNING
          —Forsyth Independent
          May 14, 1964

On October 9 and 10, 1964, 133 people gathered in Butte, Montana, to talk about the future of Montana's coal. The first Montana Coal Resources Symposium, as the meeting was called, convened in a hotel in downtown Butte, and when all the talking was over visitors from other areas were treated to tours of Anaconda Company's Berkeley pit, the huge hole in the hill that was even now gnawing its way into the city itself, consuming buildings, streets, and private property in its insatiable hunger for low-grade copper ore. What the field trips had to do with the future of Montana's coal is obscure, except, perhaps, as some kind of unfortunate omen that no one would have recognized at the time.

The symposium was attended by, among others, a group of students from the Montana School of Mines, a powerful contingent of Anaconda Company executives, the mayor of Butte, the mayor of Roundup, various coal-mining executives, a rancher from Birney, and five representatives of the Montana Power Company. They heard seventeen speakers give an almost stridently positive view of the opportunities for the development of the state's coal. This optimism was based on the observation made by one of the speakers that 68 percent of the United States' recoverable fuel reserve was coal, that other fuel sources were being used up, and that Montana itself contained about 13 per-

48

cent of that coal. One of the speakers, Dr. Arnold Silverman, a professor at the University of Montana, promised that this coal was becoming more valuable to industry as mining efficiency was improved.

"The rapid development of improved mining techniques and equipment, such as hydraulic mining, larger bucket and dragline capacities, and lower-cost, more efficient equipment maintenance and handling, will continue to increase worker productivity, already the highest in the world by far," he said. "Efforts to reduce transportation cost have led to the development of pipelines that can move coal-water slurries quickly, efficiently, and cheaply. The railroads, in order to meet this transportation revolution by cutting costs, have introduced newly designed coal cars, the unit and integral train, and faster loading and unloading facilities. The development of extra-high-voltage transmission from mine-mouth generating plants will also cut the transportation cost burden to the operator."

Another of the speakers at the symposium was Paul Schmechel, an executive assistant for Montana Power Company. Schmechel, a solid, powerful ex-naval officer, was the last speaker on the first day. His topic was "The Competitive Future of Coal."

Schmechel was, in a steady, unemotional executive sort of way, passionate about coal. He looked like a man who was about to break down a door with his shoulder, and the door at that time was the one that opened into the United States coal market.

"In the Rocky Mountain coal area states," he told the group, which had thinned somewhat from the morning crowd, "we seemingly have all of the assets necessary to enter the competitive arena. . . . We have tremendous quantities of strippable coal that can be moved by unit trains. . . . The coal industry should have no difficulty in meeting the challenge."

Schmechel concluded with a burst of optimism that must have sounded a little odd coming from a man whose company owned a coal mine that had packed up its shovels and expired six years before: "Of one thing we can be sure," he said. "The coal industry will not roll over and play dead whenever it is faced with a new threat. It is going to be aggressive, as it has been throughout history, and it is looking forward to a great future."

By that time it was known in the industry that Montana Power was planning to build a 180-megawatt coal-fired power plant in Billings. But that was almost of the scale of a toy compared to what was being planned elsewhere in the West. As Montana was discussing its coal future in symposia, the largest group of public and private utilities ever to gather under the umbrella of a joint project organized a group called WEST (Western Energy Supply and Transmission Associates). This consortium announced to the world in 1964 that it would build power generation projects in the Southwest totaling about 36,000 megawatts. *Time* magazine, groping for some way to express this vast figure in understandable terms, found an odd equivalency: This quantity, it said, was about as much power as would be generated by seventeen Aswan High Dams. The West was going to be inundated by power plants.

The first project in this vast effort, WEST announced, would be a 750-megawatt plant to be located near where the states of New Mexico, Arizona, Utah, and Colorado meet in the place called Four Corners. The plant would be built near the sleepy western town of Farmington, New Mexico. Most important, the plant would burn coal mined not far away. The power generated would flow to Los Angeles, Flagstaff, and Phoenix, hundreds of miles from the site of the plant. The Four Corners plant became a pioneer in the concept known as mine-mouth generation, in which the power, rather than the fuel, is transported long distances and the problems associated with the plants, such as pollution and industrial clutter, are visited upon a few rural residents rather than the people who use the electricity. The concept appealed to both utility executives and urban residents. The only people who would object were the rural people and they, of course, had insignificant power to interfere.

13

The middle of November she started,
Friday the thirteenth was the day.
We welcomed the moisture the first storm brought,
But figured winter had not come to stay.

But it lasted, grew colder, the wind blew
The Mercury took a nose dive
It lasted, we fought it for four months
Trying to keep those cattle alive.

—Wally McRae, from
"The Winter of '64"

Hidden from the storm behind a window crusted with frost, Wally McRae was playing cribbage with Red Kluver at the big kitchen table at the Kluver Ranch down the Rosebud. The wind did not shake the old building as it piled more snow along the fences and in the yard, but it thundered distantly like an endless train. Occasionally Kluver scraped the frost off the window to make a small porthole and peered through at a thermometer mounted outside. Five below.

It was January 1965. The snow had been smothering the Great Plains since early November. Drifts, tops curled with the wind like huge frozen combers, seemed to fill in the wrinkles of the land until it was all just a sea of white, out of which the red cliffs, pale now, seamed with snow, rose like islands. There had been no respite, no chinook wind to break the frost. Cattle stood in clusters, their red fur clotted with chunks of dirty white where the snow had blown against them and frozen. Some of the cows had already died, to become bloated hard lumps swiftly covered by the snow that, even on clear days, swirled like spume across the surface of the land.

In some parts of the West military helicopters, of the same design that were becoming familiar on television as the ones used in Vietnam, had airlifted hay to stranded herds.

McRae, who fed his thirty head of bulls in a pasture not far from Kluver's place, had stolen this hour to visit out of a day which had already worn him down another notch in his endurance race with winter. He had come mostly for a little warmth for his fingers and cheeks and a little companionship, but also to touch base somehow with longer experience. This winter was worse than any in his memory, but how bad was it, really? He remembered all the stories his parents told him about the depression, as if there was a lesson there for him, and now he wanted to know what it was.

How much *can* we take? How long before we *must* crumble, sell the carcasses, and go to work for wages? Perhaps he sensed now, as an historian would theorize much later, that the vast West of mountain and plain and violent weather is a particularly demanding crucible of natural selection, sorting with the most vehemence through the membership of the human species. How do I know when to quit?

But McRae had learned, in the years since he branded with his father in 1944, that not only do you never ask how many calves your neighbor has or suggest improvements in the way he runs his business, but you also do not bare your own troubles to his sympathy. This community of ranchers along the Rosebud Creek is remarkably close-knit in spite of the distances that separate neighbors—or perhaps because of them—but it is a conservative community, based on tradition, and much of that tradition is of restraint. You did not ask your best friend what he left behind in Texas when he came up the trail; it might have been a corpse.

So McRae stalked his question like a big cat, a rancher prowling on sensitive ground. They drank coffee, bantered about the game, and laughed about the weather. Red retold an old story: " 'Never seen it so clear and still,' sez the old-timer in the blizzard. 'Snow *clear* up to my rump, and wind *still* a blowin'.' "

The year had not been an easy one on the Rocker Six. Wally got his leg out of the cast in June, but it came out emaciated and weak and took a long time strengthening. Hired hands came and went while Wally searched for the one who had the right blend of enthusiasm and experience to create the all-important relationship of mutual respect. There were a few high moments: Ruth, it turned out, had only lost one child of a set of twins. In June, Allison was born. McRae had quickly begun to fill his father's large boots in the community; he had become active in the Montana Stockgrowers Association and was already developing a reputation for hustling memberships. One of his ambitions was to become a director of the organization. In a community based on tradition he had not stepped out of line with the traditional conservative Republican politics: Like almost everyone else he had supported and voted for Senator Barry Goldwater in his disastrous race for president against Lyndon Johnson. And in his limited spare time he had begun to pick up a few extra dollars announcing at rodeos in Forsyth, Miles City, Hardin, and Billings.

He had earlier satisfied the ham in him by performing as a rodeo clown, distracting bulls with antics. He once almost caused the suicide of a rodeo manager who had hired him, through a friend, for the verbally demanding announcing job. Meeting the harried manager for the first time fifteen minutes before the rodeo began, McRae stuck out his hand and smiled.

"My name is Wally McRae," he said. "G–g–g–glad t–to be workin' here t–t–t–today."

He had also begun to find a niche in the Rosebud community for his poetry, lighthearted rhymes about ranch life that amused his neighbors and were occasionally published on Production Credit Association calendars:

> I've quaffed some beer from a frosted mug
> And rum from the tropic isles.
> I've sipped champagne at the Country Club
> In the cowtown known as "Miles."
>
> Red Chianti from a grass-wrapped jug
> Passed my lips in Italy.
> I've wintered on Brazilian coffee.
> Summered on Ceylon iced tea.
>
> But the best damn drink I ever had,
> Came from a green quagmire;
> Sipped, face-down, in the August sun
> After a prairie fire.

But most of the year had been difficult. With the two loans hanging over him, McRae had established an operating budget of just over $800 a month, which included $325 for a hired hand, $150 for his family, and all expenses except feed and loan payments. It was a thin budget for a ranch his size. He could hardly afford essentials. But he had calculated survival on the basis of that budget; what he hadn't figured was that his first winter would be his hardest.

The first storm, in early November, caught him with cattle scattered and calves unsold. Some died while he waited for the normal following thaw. But by Christmas it became apparent that warmer weather would be too late if it came at all. So on Christmas Eve McRae had a narrow road plowed to one of the distant pastures and trailed many of the cattle in. It was a disastrous day. He had to work his horse nearly to death keeping cows out of the worst drifts, and he could do nothing to help when, one by one, the weakest ones stumbled, dragged along on their knees a few steps, and then collapsed. He had to leave them in the snow, and the herd, like soldiers in retreat, filled in the gaps and struggled on.

And even now, while he sat in Kluver's kitchen, the storm continued. The wind had picked up new vigor from the west, blowing at five below with enough force to exhaust the lungs of the Arctic. Plumes of snow curled off the ridges and off every drift, and there were two new dead cows lying stiff-legged in the pasture. Others would not last much longer: As McRae fed them the daily ration of hay and cow-cake they just stood there, uninterested, ribs standing out like the marks of a lash, backs humped, ears and tails frozen, waiting in a stupor for the end. Each death, when you thought of the feed, and the time, and the unborn calf, was maybe $250 gone, but it was not really the money. Each death was a failure—of judgment, perhaps, or of resolve.

So the game of cribbage between hours of work in the storm was not all recreation. Kluver and McRae laughed and drank coffee, this big, old hard-eyed man, with sandy red hair and a face freckled by sun, and this younger one with the heavy moustache and the small, flashing eyes. They talked about good bulls, about the calving season coming up in late March, about a fellow named Ambrose who used to operate a small coal mine nearby and who, the story had it, once blew up his dog instead of the coal seam by mistake, and about McRae's two children. And all the time McRae was wondering: Can we take much more of this winter, Red? What if it snows until June? But the questions were too pointed, too much an admission. So he hunted around the perimeter, and talked casually about dead cows.

Finally there was a pause. McRae lit a cigarette, striking the match with his thumbnail. Kluver leaned to the window and scratched another porthole. Still five below. Then McRae's words slipped out, close as he could get to his concern.

"Yeah," he said lightly. "Sure is tough."

"Yeah," Kluver replied mildly, his eyes bright with some ferocity that his tone did not reveal. *"So are we."*

As McRae drove back up the road toward home, the snow blowing up behind his pickup in an opaque white cloud, he was caught by an exhilaration, a sense of being swept up in combat.

"All right," he remembered thinking. "That's all I needed to know. Now I know what he's talking about. Now I know what I gotta *do*. All right, winter, damnit. I'll fight you. You just keep dumping that snow and that cold on me and I'll just back off and I'll just fight you and we'll see who's still around here in the spring."

The blue flatbed truck thundered on up the scoria road, and McRae left behind him the only time, in all his struggles with nature, cattle, corporations, his weak left leg, and the power created by coal, that he ever seriously considered giving up.

14

To the girls in the telephone building in Havre, Montana, the men who appeared outside their high windows one day in late 1963 on a slender framework of brown steel were hardly real. One day there was space out there—broad sky and emptiness, and the next, it seemed, here were men walking outside the window on those metal twigs like sudden handsome birds.

The height stole propriety. Everything seemed suspended in the bright air through which the men walked, riveting together the beams that came sweeping slowly toward them on slender cables; everything seemed suspended: responsibility, time, reality. So the girls, demure on the street, would gaze brazenly out at the muscled figures balancing on metal ledges six inches wide, thirty feet from the ground. The girls wore their hair in short sweeps and bunches, and their lipstick was bright red. From the steel framework of the new building their lips bobbed behind the window glass like apples.

Vic Jungers was an ironworker. He was a connector on the crew of men who were assembling the first beams of the new building in Havre. They were bolting a structure together out of a stack of steel bars. He was a connector and not without pride; connecting the iron is the best and most demanding job in the ironworker's craft; it demands balance, agility, and a kind of confident fearlessness that disdains height. You walk on the first tentative thrust of the building into the air; you straddle the no-man's land between emptiness and solid form. Structure congeals beneath you but up there it is just wind and steel. Jungers loved it. His father was an ironworker; now he was an ironworker. The time was yet to come when he would meet the woman who would tell him an ironworker was just a man with a suitcase full of dirty clothes and a hard-on, and he would be forced to believe it. He walked with a slight swagger on the high beams, and he knew that the girls in the telephone building were watching.

But there was a restlessness in Jungers too, which didn't go with the self-confidence of the man of high steel. Jungers had tried to hang himself when he was a child. When he became an adult he liked to say he was disappointed he didn't succeed. "I missed a great opportunity," he would say, as always sardonic and slightly removed by humor from even that personal reality, the way he walked slightly above the world on the tops of new buildings. "The only reason I've never been able to go through with it is that I know what's here and I've never talked to anybody on the other side." And that insecurity remained with him even as he rejoiced in his agility and balance on the iron. "I've always wanted to be a great person in one way or another," he said much later, "and I realize I probably never will." It nagged at him, the perception that he had more potential than he was using, even in 1963, when he was twenty-five.

One day that year he climbed to the top of the structure, to the open framework just one story above the place where the floor was already in place and where carpenters were beginning to flesh out the building. It was one of those bright, cold days in Montana when the sky looks as if it has just been scoured by a storm. Jungers climbed a column to the top and walked along a beam. To him it felt two feet wide, like a bridge, but in fact it was only wide enough for one foot. He strolled along it, his angular face concentrated on the work ahead of him, but his mind tingling at the presence, in a window, of the figure of a girl. She was just a bright form on the edge of his vision, but her body was shaped to perfection by his peripheral imagination. She moved, her lips danced, and Jungers fell backward off the beam.

In the long gasp between floors, before he hit the planking and felt the pain, Jungers had a lot of time. Perhaps it was then he decided to go to college. And when he finally went away to school a few months later, packing up his union card and heading first for Wyoming and then for San Francisco State College to become a teacher, he never thought he would work the high iron again.

15

Colstrip is to furnish fuel for Billings' new Montana Power steam-electric plant. The 180,000-kilowatt plant . . . when completed will be the largest coal fired plant in Montana. . . .

—*Forsyth Independent*
July 7, 1966

It was the spring of 1967. The walk from Martin White's desk in a kind of hall of accountants to the office of his boss, Montana Power Company Vice-President and Treasurer J. J. Harrington, was up four floors. But White did not make the trip with the bounding enthusiasm he used to have, when he was young, six months before. He did not know what the man wanted, and perhaps he didn't care. Termination or advancement were no longer significant matters.

Martin White had originally planned to go into the U.S. Forest Service as a career. But his father talked him out of it: "Too much politics in the bureaucracy, kid." So in the spring of 1966, after graduation from Montana State University in math and economics, he got married, checked in 1A with his draft board, and went to work for Montana Power Company as a clerk in Butte. It was a job with little promise as a career and no politics whatsoever. "I'll just take this job and bide my time till I get drafted," he thought.

With the frankness of expression that comes easily to a man who expects to be shipped off to war any minute—and to the son of Stewart White—Martin White did not hesitate to criticize the accounting procedures of the company that hired him, at least to his fellow workers. His comments did not endear him to the older staff.

"I was one of those rebel guys," he remembered later. "And I didn't understand why all these things were going on. Jesus, I used to get some of the most interesting lectures from the old guys in the office, why I shouldn't question what the company was doing. Their loyalty—I couldn't get over it—they were fiercely loyal. They told me you'll never go anyplace, you'll never go *anyplace* feeling like that."

Then, on November 14, 1966, his mother had called from a pay phone. He could hardly hear her; he thought it was a bad line.

"Your father," he thought she said. "Your father is under the truck. At Nineteen Mile."

White, who had bought his own four-wheel-drive truck sometime earlier, left work and raced out to where his mother was waiting. He had visions of the frank, outspoken man he was so fond of, lying pinned under wreckage, in pain. But when he met her he found that the news was different: he had misunderstood the message. While the Whites had been hunting rocks in the mountains, the truck had become stuck. While trying to dig it out Stewart White had suffered a heart attack, and had died. Martin's mother had left him and walked out for help.

"He'd a wanted it that way," Martin remembered. "He wasn't the sort of person that'd a wanted to linger, and he was really an avid guy. But it was a hell of a thing."

But there was more to come. A few weeks later Martin, who had joined the Reserves, got his orders to report to basic training. And while he was turning his energy into an appointment as a platoon leader at Fort Lewis, his new marriage was falling apart. "So it wasn't long after I came back that I got divorced.

"I was—I just never—our family was so happy that a divorce was something that always happened to other people," he said later. "It was tough. And after I got divorced I kind of got wild there for a while."

By the spring of 1967 White was more dependent. on Montana Power than he had ever expected. It was the one constant in his life. After his mother moved away to live with her mother, nothing else in Butte gave him an anchor but his work, the regular hours in the tall building at 40 East Broadway, safely distant from the Anaconda pit into which Butte was falling, grain by grain. So when White sat down in Harrington's office, he was not the same rebellious young man he had been in the fall. And the treasurer, either thoughtful or cunning, recognized his power to give White the handle he needed, the place to grip.

"Dig into your work," Harrington said. "Work like hell. And one of these days it will all be by you and at least you won't have sacrificed this work period."

White listened, nodded, and walked back downstairs.

And one day when the department supervisor came through the accounting department, asking casually if anyone would like to take on, as a small extra duty, the books of a relatively insignificant new subsidiary, Martin White volunteered.

At the time White was handling the books for another Montana Power subsidiary which was involved in natural gas development. When the supervisor came through the room offering the books of a little unit called Western Energy, which took care of the company's coal property, there was a subdued mumble from the accountants. *Another* little extra. He *says* it's a short book. Who's the goat to get yoked to this one? Sorry, I'm already swamped. Don't look at me; I gotta see the wife and kids at least once a month. So when White, the handsome, young, blond, blue-eyed boy of the office, put up his hand, in effect, to take the extra work, he must have looked insufferably eager. It was not unlike him to be enthusiastic, but this time, at least, there were other motivations. The growing work load was like the sound of a band at a funeral; the louder it got the more it covered the tears. He welcomed its demanding rhythm. It carried him through.

But there was a price. "I think he went through something that was really awful in those years," his sister, Sylvia, said later. "Because he really got hard-jawed after that. He still gets really excited about things, like his hunting. But I'd say he got tough after that, not tough with other people; but you just can't get into Martin's feelings anymore."

16

The town of Colstrip may not see the staggering output of its heyday, but there is now no question but what operations on a relatively high scale will once again be seen by the end of summer.

It is estimated that by 1975, 60 men will be employed at the Colstrip operation which will supply about two million tons of coal annually. . . .

—*Forsyth Independent*
May 16 and July 4, 1968

The twenty-four-car Burlington Northern train rumbled north, down the tracks beside Armell's Creek, through fields still green, yellow, and blue with the spring's grass and flowers. The day was hot and still, and the cattle gathered in short pools of shade on the hillsides. The clatter did not disturb them; they were more concerned with the flies. Inside the elegant business car that was tacked absurdly on the end of a train full of coal like an orchid pinned on overalls, Uuno M. Sahinen of the Montana Bureau of Mines, a man who had attended the First Montana Coal Symposium four years before, turned to a reporter from the *Billings Gazette.*

The public, he remarked, really shouldn't complain about mining companies that, in their words, "despoil the land." After all, he said, less than twenty-five square miles of this large state had been dug up by mining.

"And," he added, "if they really want to quibble about it—look at that landscape over there." He gestured toward the red-topped buttes, fringed with Ponderosa pine trees, marching back into the distance in jagged rows. "According to some standards a mining company who left land looking like that could be prosecuted."

61

The two men laughed.

It was July 5, 1968. This was Montana Power Company's hour of triumph: the first trainload of coal delivered from the reopened strip mines in Colstrip to the brand new 180-megawatt Corrette Power Plant in Billings.

The business car was filled with an air of festivity and a steadily thickening odor of alcohol. The carload of mining officials and reporters had just been on a guided tour of the Western Energy operation at Colstrip. They had seen the rejuvenated twenty-four-cubic-yard shovel used to carve chunks of shale from the roof of the coal; they had seen the black seam lying, like a river of darkness, in the trench where Northern Pacific had left it; they had watched the old town bustle with its new importance; and they had heard bright predictions.

"I don't know anything that's of more significance to this area than what we're seeing now," said the editor of the *Miles City Star*. And Robert J. Labrie, assistant chief engineer for generation for Montana Power, confirmed reports that the company was making plans for another coal-fired plant, perhaps twice as big, but that its location had not been decided.

And now the load of dignitaries was celebrating its way west, with coal rolling ahead of it and drinks on the house. It was a buoyant moment: The future stretched out ahead of the company like the straight track to Billings, laid on a bed of coal. No longer would Montana Power be forced to fight the vagaries of rainfall and the moods of the federal government and Indian tribes to generate its power or to build new facilities. Back in 1966 the company's annual report had promised this moment, announcing its newfound freedom from two proposed but often postponed hydroelectric projects:

"The timing of both the High Mountain Sheep and Buffalo Rapids developments is uncertain," it had said. "But your Company is in a position to build its own steam-electric generating plants as required to meet load growth." This train trip was that promise fulfilled. Montana Power, like most other utilities in the western half of the United States, had discovered the rich bed of coal in its backyard. The only uncertainties left in the way of unlimited generation were just problems of engineering.

There were no doubters on the train; or if there were they kept their questions to themselves. Even the reporters seemed caught up in the enthusiasm. In the following days the *Billings Gazette* printed several articles about the coal and a glowing editorial.

Pointing out that Wyoming coal development was ahead of Montana's, the *Gazette* suggested that "Montana, if it wants to get its fair share of development, is going to have to provide the same kind of industrial incentives and encouragement that Wyoming provides." The paper further praised the company: "Montana Power Company, with its Billings plant, and its interest in encouraging other uses for coal, has made a fine beginning."

Only two small irritations marred the bright picture. In an analysis of the event, reporter Dick Gullily expanded on Sahinen's comment: "Conservationists ... are critical of the 'spoil banks' at the coal mine ..." he wrote. "[Montana Power] replies that the land is marginal anyway." The only other minor criticism, he added, was that "many conservationists claim the company should never have built the plant near a populated area."

But these were just undertones against a great clamor of success.

"I think you'd have to say that was a high point," George O'Connor, who by now was president of Montana Power, said later. "It was the first pound of coal the company had ever shipped. And it was not only the first coal we'd shipped but it was a new approach to generation. From our standpoint this was a breakthrough. And we were rather proud that we had taken what we thought was a pretty advanced and pretty adventurous step."

O'Connor added that by the time the train reached Billings the celebrations had been so vigorous that while the reception committee needed a chute to unload the coal it needed a steam shovel to unload the passengers.

Martin White was not on the first coal train, although his boss, Paul Schmechel, was. White was too busy in Butte helping to organize the operation that made it possible. Because since the day that he had decided to accept the minor duties of handling the books of Western Energy Company, he had been up to his eyebrows in coal.

17

*Peabody Coal Company will open a new coal mine just south of Colstrip. . . . Op-
erations at this mine will commence shortly with coal delivery to begin by early
1969. The coal will be sold to Minnesota Light and Power Company. The expected usage will be 2,000,000 tons by 1972.*

—*Forsyth Independent*
August 8, 1968

The lawn on the commons, the wide circle of grass in the center of the
campus of San Francisco State College, was slowly being trampled to
death. The grass was scuffed and gray, and littered with paper. In the
quiet evenings it almost seemed as if there had been a big picnic there,
until you picked up one of the scraps and read it. "This University Belongs to the Student. Dig it!" The bits of paper were all pamphlets and
fliers stamped with slogans and line sketches of clenched fists. Each
night the janitors gathered them up, and each day they were replaced.
It was not quite the same with the windows on the flat-faced buildings.
When they were broken by rocks the janitors picked up the glass and
replaced it with plywood. Across the campus, each day was similar, a
cycle of noon anger and evening peace, but the mood of crisis never
died—like the blood on the steps of the library where a man had been
injured by police, it was mopped up daily but the stain remained.

Vic Jungers was out on the commons every day. The ex-iron-
worker, now almost thirty, was studying American Literature, still
planning to teach. But his labor background and his efforts to organize
a union in the cafeteria, where he worked, had led him to a loose association with the group called Students for a Democratic Society (SDS),
the most radical organization on campus. He was something of a fringe

participant: Both the organization and he had certain reservations. SDS frowned on drug use, and Jungers enjoyed the journeys of his mind—he once told a friend at the university that he would be perfectly happy to vegetate in bed all his life, with glucose fed into one arm and LSD into the other. "It's a head trip, life is for me." And Jungers, who believed in nonviolence as much as he did in the pleasures of chemically induced states of consciousness, did not agree with the SDS tactic of creating confrontations between students and police every day at noon.

But each found the other useful. To SDS Jungers was a hard worker and a clever ideologue. He collected bail money, wore a red armband, and argued passionately for the cause: the acceptance of fifteen Black Students' Union (BSU) demands; the reinstatement of a professor who had been fired after he allegedly urged black students to bring guns to class; and the defiance of S. I. Hayakawa, the acting president. To Jungers SDS was a receptacle into which to pour his intellectual restlessness, where he could mix it with his idealism and come out with some kind of hopeful broth that would have just a hint of greatness. "I was naive enough," he remembered, "to believe that I was going to get something out of this bugger for the white people, too."

The illegal noon assemblies on the commons were curious episodes. Everything seemed orchestrated, with only one instrument—human anger—playing wildly out of tune. The police were practically bivouacked on campus; with the noon hour both they and an assortment of BSU and SDS speakers emerged, and set up confrontation like a handful of stagehands preparing a scene—rostrum on the chalk marks, hecklers center stage, newspaper photographers to the left, riot sticks at port arms. There would be more spectators than participants lining the lawn, wary and a little jumpy when the San Francisco Tactical Squad came out. This organization of burly men, safe behind plastic visors and helmets, bearing pale yellow sticks, put more spice in the afternoon than any revolutionary rhetoric from the speakers' platform. The Tac Squad's method of advance, in which the two ranks of men would burst forward with a flurry of sticks, to halt at three paces, sent the crowds swirling backward like a wave off a seawall, only to wash back in, close as they dared, with taunts and snickers, until the next advance.

Then on some days a stone or a bottle would fly out of the crowd, and when the Tac Squad rushed forward it would not stop.

Jungers was in the cafeteria, an old building serving more people now than it was designed to hold. He was collecting bail money in a coffee can beside a relish table. Ketchup, mustard, little bags of sugar. Jungers was sitting near the table. He remembers the scene the way an accident victim may remember the color of trees, the faces of pedestrians at the moment of the collision. As he reached to take a quarter, there were screams in the crowded room. A friend of his, a member of SDS named Paul, darted through the crowd; behind him came the squad.

"I jumped up to see what was going on," Jungers remembered much later. "I was right at the end of the table. Long tables. Steel chairs. And I stood there and Paul run by me, fifteen feet, into the dish room. Two cops went in, one came out. Pretty soon here come this cop out and he had the billy club and he was choking Paul down with it.

"Paul was looking at me and I could tell by the look in his eyes, he was pleading with me to help him. And I'm standing right there lookin' at him. That's what I felt at the moment."

Jungers hesitated. All his principles were on the line there, opposed: nonviolence or revolution; peaceful dissent or hatred of the pigs, the lackeys of white supremacy; restrained idealism or friendship. Paul's eyes, asking for help.

"Maybe I made a move, or something. I don't know. I was thinkin' it. I definitely admit I was thinkin' it. All of a sudden whammo! My knees gave out. I went down and I grabbed the table. I reached up and grabbed my head where the pain was and the blood was just gushin' out of it and this cop walked by and I was infuriated, and I picked up a chair and I hollered at him and he turned around and I laid him out on the floor.

"Then I took the chair and went to work on the cop that had a hold of Paul. I went wild.

"And we both split. Into the washroom, underneath the wash rack thing and down the back into the linen. We were in there for about five hours."

After escaping from the linen, Jungers went home, told his personal physician that he had fallen down the stairs, received fifteen stitches, then saw on the evening news that a policeman had been hit by a chair and was not expected to live.

Murder. The news washed away his anger like a cold shower. The events of the afternoon became twisted, spun into endless repetition by a centrifugal whirl of fear; fear more of having committed the unforgivable act than of capture. The ketchup, the mustard, the blood, the blast of hate, the swung chair. But did it have murder in it, that hate? Murder from a man who did not believe in violence? What destruction would it bring if he had killed a man? Again: the running friend, the screams in the crowd, the flail of yellow sticks, the shot of pain, the swung chair. The swung chair. *The swung chair.* Bent on the head of a man. Of another man. Death in his hands, the snap of vertebrae. Who was he to bring that sudden darkness? Again: the coffee can of change, the food, the confusion, the look in Paul's eyes, the swung chair. The steel chair. The chair.

"For two days I sat there; didn't go to school: What happens if this man dies? I didn't know for two days that he had been hit outside the cafeteria, not inside. It wasn't me."

In the spring of 1969 Vic Jungers moved to Rosebud County. He came back to ironworking, although he would not entirely forsake the academic life for another two years. He came to Colstrip to help construct a building at the Peabody Coal Company mine about two miles down the road from Montana Power's Western Energy operation, and he came to get away from San Francisco State. He rented a house in Forsyth and commuted. No one bothered him; no one intruded upon his brooding summer. Right there in his front yard, on Main Street, he grew a small patch of marijuana. It was the first such crop in Rosebud County.

18

Another contract for Colstrip coal which will practically double the current production has been obtained by Western Energy Company, it was made known last week. The shipment of 110 carloads a week to the Wisconsin Electric Power Company at Oak Creek, Wisconsin [will begin] a trial for 15 weeks....

—Forsyth Independent
November 28, 1968

In May 1969, the Montana Stockgrowers Association, an organization which had, during this tumultuous decade, nailed down its conservative viewpoint by vigorously resolving "that our association collectively and as individuals rededicate life and energy to the Great American Heritage," held its eighty-fifth annual convention in the cow town of Miles City, fifty miles east of Forsyth.

The nine hundred delegates heard a keynote speaker point out wryly that cattle prices during the first four months of 1969 were below the prices achieved in 1951; they looked over promotional ads for their product—"Beef for Father's Day: Make Pappy Happy"—they were solicited themselves—"Purina Range Cattle Spray for Cows," "Triple P Vaccine for Calves." They made fun and music—a photo in the association's magazine for that month shows "Doc Forsberg and Wallie McRae" singing before a microphone.

McRae came to the convention with two ambitions. One was to be elected a director of the association, a position he had coveted most of his life. The other had to do with coal. Because by the time he came to Miles City McRae was no longer innocent of the changes that were coming to his county.

About a year before three men had come down to the Rocker Six looking for Wally McRae. They had been wearing suits and ties and they looked overdressed sitting on the steerhide on the couch in McRae's living room. They were very friendly; they drank Ruth's coffee and made compliments. McRae watched them carefully out of his narrowed eyes.

They worked for Peabody Coal Company, they said, and they were interested in surveying his land for coal. Federal homestead laws, after 1909, had given settlers surface ownership but had retained the coal and the iron under the homesteader's scab of surface. The railroads had earlier received huge land grants from the government, and although some of this surface was later sold, the railroads also retained the coal and iron. So even if the rancher owned a block of land surface, the coal under his land was, in most cases, owned in a checkerboard pattern by the federal government and the railroads. And Montana law, at the time the men came to visit, still allowed mining companies to condemn private property. So McRae felt like a rodeo cowboy with one foot caught in his stirrup and one arm wrapped around the horns of a steer.

The three men made quite a point of asking McRae for permission to come on his land and drill prospect holes. They were so polite about it that McRae became suspicious.

You might as well know if you've got coal, Wally. Might be worth something to you. Your neighbors are letting us drill, and it looks like the seam goes right through your land.

Perhaps out of instinctive irascibility, or perhaps because his sense of Montana history reminded him that cattlemen and miners had been fighting for political control in the state since it was settled, McRae told the men he didn't like the idea. "I run a ranch, not a mine." The line hardened.

Well, Wally, you don't own the coal, you know. The railroad owns it, every other section, as spoils for pushing west. We've got the right to prospect. Of course, let's visit about it, let's be reasonable. You might get a well or two out of the deal.

Before giving permission McRae called his lawyer and the lawyer said you've got to let them on. Okay, McRae told them, but I don't want any wells, you damn well better not damage my land, and don't use the water from my reservoirs.

Then one warm summer afternoon, Rip Sept, a hired hand, came to McRae with a report: "Somebody dropped the oil out of their outfit and threw a bunch of oil filters over in the summer pasture," Sept said. "And you can see the tracks where they backed into the reservoir over there at the head of the Shaver draw. You can see their tracks and the print of the hose."

"I think," McRae remembered, "I think that's the first time I got mad."

Several months later, after McRae had thought the core drillers had finished, a light blue Ford pickup drove in off the road, over the rattling bridge, up through the yard where the seasonal equipment stands waiting for the year to turn, past the line of juniper trees planted years ago, and out onto the horse pasture and down over a ridge. In ranching country you identify people by their vehicles, but this was kind of a ghost, an unfamiliar and fleeting vision that carried itself with an odd authority, barging through where even a friend would have stopped.

The second day it came McRae was there to meet it. He didn't have a gun, but his hard eyes were weapon enough.

"What are you doing on my outfit?"

"I'm surveying for Peabody Coal Company." The man spoke a bit wearily, tired of straightening out misunderstandings for irritable ranch hands.

"I thought you folks were through."

"No, we've got all kinds of holes to poke down out there." The man grinned, and showed him a speckled map.

"Well, I'm getting a little tired of you guys thinking you own the outfit," McRae said.

"Take it easy, buddy," the surveyor said, in his best soothing voice, which came out sarcastic. "I've just got a job to do and I'm just doing it."

"Well," McRae said, "I haven't seen one of you birds for six months, nobody has contacted me, and as far as I'm concerned you're trespassing."

The surveyor unlimbered his trump card in exasperation.

"Okay, look. I've got permission."

"Who gave it to you? What's his name?"

"Wally McRae."

McRae paused. So. There is nothing so pitiful as a man whose last bluff has no legs. This would be a pleasure.

"Oh, no he didn't," he said.

"Oh, yes he did."

"No. He didn't." McRae said, stringing the game out, arms folded, face a thin-eyed mask.

"How do you know?" the surveyor said, growing nervous. There was a trace of humor in McRae's face, and that made it all the more threatening.

"Because I'm him."

Trying not to grimace, the surveyor surreptitiously put the truck into gear. But his mind hadn't caught up to the situation. He became bombastic.

"The hell you are!" he said.

McRae ran him off. But as the man turned the truck around on the lumpy ground he was angry.

"I'll go," he said. "But I'll tell you one thing, cowboy. I'm coming back."

And somewhere in McRae's mind a line was drawn, a hard, deep scratch in the earth of Rosebud County between *us* and *them*, with the people who wanted the coal dug from the ground banished to the other side. It may have been the tactics more than the general goal that first offended McRae, finely tuned as he was to his image of the code of honor of the West. But his decision was permanent, and eventually it would encompass a raging opposition not only to the early rudeness and carelessness that the coal men brought to his county, but also to everything they did to the earth, the people, and the air, culminating in that roaring expression of all those things combined, the power plant. But that was still three years away. When McRae went to the stockgrowers' meeting in Miles City he carried with him a proposed resolution aimed at just the first symptom of the issue, the mining itself.

Since the stockgrowers are a powerful interest group in Montana, their meetings are attended by representatives of most government agencies and many other groups with something to gain or lose by their actions. It is a tradition to introduce these visitors early in the general session; one year there were so many tied and jacketed men with tans clear to their scalps bobbing up and nodding in the audience like painted corks that one old cowboy finally stood up and asked, "Is there anybody here that runs a cow?"

One of the visitors in 1969 was a man named Sid Groff, who worked for the Montana Bureau of Mines and Geology. Groff had been promoting coal development in Montana since 1961; he had been the general chairman of the First Montana Coal Resources Symposium at which Paul Schmechel had spoken in 1964. The only possible restraints on coal development in Montana he could even imagine on the horizon were tangibles like stripping efficiency, BTU content, and market conditions. So he was stunned when McRae stood up at the convention and proposed that the stockgrowers officially urge the state legislature to pass a law regulating strip mining.

"Whereas," his proposed resolution read, "the present practice of leaving spoil banks not only destroys the scenic natural beauty of the land, but leaves it unsuitable for the production of livestock,

"Be it resolved that the Montana Stockgrowers Association recommends that stringent strip-mining laws be enacted by the Montana legislature and that these laws be strictly enforced."

When Groff objected to the proposed resolution the chairman of the meeting sent him, McRae, and three others off into committee to wrangle it out while the rest of the convention turned its attention to a less strange request: a resolution to urge legislation to curb meat imports.

Outside the hall Groff turned to McRae. His words stuck in McRae's mind: "I couldn't tell you inside there," he said. "But if you think you're going to obstruct coal development in this state, you're crazy."

Groff, McRae remembers, was angry. In the small, stuffy room where they were sent to compromise, Groff must have felt frustrated at the threat of a new block being thrown in front of coal development, however advisory and without muscle it might be. He must have sensed in the resolution some seed of danger, threatening to undermine the new prosperity that still seemed tenuous. But he let his anger out, and it defeated him.

"There's nobody going to stand in the way of progress here," he said. "We're going to mine this coal and we're going to mine a lot of it and we're going to create a lot of jobs." It must have seemed absurd to Groff, who spent hours developing scholarly game plans for the development of Montana's coal, scenarios weaving water, projected energy demand, seam thickness, and overburden depth into predictions of economic growth, to be arguing in this back-room closet over whether or not land that was nearly worthless to begin with should be re-claimed at great expense. Why, he had said so many times, the total amount of strip-mined land in the state probably would never exceed two hundred thousand acres, just a shave off the great mass of Mon-tana, and to put some kind of forced reclamation requirement on com-panies in place of the state's voluntary programs would do little but stall development.

"To be perfectly frank," he said, "nobody cares what you cowboys do. There just isn't going to be any restrictive legislation placed on coal development."

McRae's eyes pinched down to needles. He had been nervous in the big room. Now he was relaxed, a cigarette smoldering in the ash-tray beside him.

"Well, Sid," he said innocently, throwing the role of unsophisti-cated cowboy that Groff had just handed him loosely around his shoul-ders. "That's just fine, I guess. Then why fight this resolution? We'll pass the damn thing. Everybody here's gonna be happy. Then they'll go up to the legislature and if what you say is true it's gonna be killed in committee. We get a good warm feeling in our heart and it's not gonna go anywhere."

"And I had him!" McRae said much later, the years of conflict and defeat that intervened not affecting his glee at the remembrance. "And so he said okay. And we passed it. And the next time the legisla-ture rolled around we passed the strip mine reclamation law."

But it was a small triumph. Because even before that action other steps had been taken that would make it of secondary importance. In September 1969, Montana Power ordered a turbine generator set for a 350-megawatt power plant from the General Electric Company. In De-cember 1969 Montana Power ordered a boiler from Combustion Engi-neering Corporation capable of burning Rosebud coal. And Montana Power began to look around for a place to put them, because by now there was political opposition to building another power plant in Bill-ings. The company needed a nice, lonely place, far from the public eye.

19

Montana Power Company, according to a news release last week, is planning the construction of a new coal-powered generating plant in this section of Montana. The report said the plant may be twice as large as the one in Billings.

The report indicated that the new plant might be east of Billings on the Yellowstone River in an area which would conveniently utilize the coal deposits at Colstrip. Montana Power, however, has made no official announcement of the plan.

—*Forsyth Independent*
November 13, 1969

"The future Denver of the Northwest," Barbara Wimer wrote in her precise, round handwriting, "is a description found in books and spoken of by many prominent men in describing my hometown of Colstrip, Montana." She paused a moment and looked up at a classroom filled with students. She bent back to the lined page, which stretched out below her first sentence like an uncharted desert.

"To most people," she wrote, paused, then wrote again: "if they ever saw Colstrip, this would seem like a perfectly ridiculous thing to say. . . ."

Barbara Wimer was an ebullient, enthusiastic girl with a face as round and pretty as the letters she rolled across the page, and the slender but strong build of a woman who has grown up with horses. She had lived all her seventeen years on her family's ranch, and had known that rare kind of stability that is no longer common even among rural children: She had attended the Colstrip schools all her life; of her graduating class of twenty-five students, ten had entered first grade with her in 1957. And like most of those ten she had been powerfully restless when the release finally came.

Her itch to leave Rosebud County had been encouraged in 1968 when, as a high school junior, she entered the district competition for high school rodeo queen and keep winning all the way up to the national finals, in Topeka, Kansas, where she finished second. Her victories had been homegrown: Her family had made the suits she wore in the arena, she had trained her own horse, Tango, and she had received coaching in public speaking from Wally McRae. But her success opened a door for her to the outside world and all through her senior year that door stood ajar while she waited through the long, slow Rosebud County winter. And when she finally graduated she packed up her riding suits, her tack, her self-confidence, and Tango and headed for Kansas once more, where she enrolled in Kansas State University, to major in business.

"When I left all I could think of was, boy, I want outta here," she said later. "I couldn't wait to get out of there. And then it started dawning on me, these things I take for granted, well, a lot of people don't even see them."

The Wimer ranch is centered on a low ridge between the drainages of Armell's Creek and the Rosebud. From the grassy hills in the divide you can see west to the Sarpys, north to the Yellowstone, south to the reservation, and east across the Rosebud into an infinity of slowly bluing, pale pink hills. There are small clear springs on the land, and windmills where the cattle drink. The place has a sense of upland to it, though it is only a hundred feet or so from the valley floors; the dropping away of the land to the east and west gives you a feeling of spaciousness. The old ranch house itself is tucked into a small cup of hills, but even from there the land seems to roll away down, surrounding the building and its attendant corrals with solitude and reach.

Perhaps in this urban maze of the university Barbara Wimer thought about those empty hills when she remembered home. But when Miss Flimmer, teacher of the English Composition I class Barbara attended every Monday, Wednesday, and Friday assigned a 250-word composition, Barbara chose to write about the rumors of change and excitement that had been stirring through the upper classes in Colstrip High School all through the previous year. While Barbara had been out racing around thousands of barrels on Tango, tying down hundreds of bleating goats, and smiling at dozens of judges, Montana Power had been mining its first coal, there were new, more worldly kids in the school, and all the talk was of boom. "It was a story-tale kind of thing, you know," she said later. "It sounded exciting at that time." So that was what she slowly and painstakingly tried to get down on the page.

". . . This would seem like a perfectly ridiculous thing to say because Colstrip only has a population of around one hundred people and as far as prosperity goes there is not even so much as one store in the town. Nonetheless, the possibility of this description of Colstrip developing into reality is becoming more feasible with each passing day as the world's former second largest strip coal mining town is beginning to operate again after a thirty-year shutdown. [Her statistics were incorrect—Colstrip had only been completely shut down for one decade.] Furthermore, to make the situation even more impressive there are two companies opening rival mines, thus competition thrives and Colstrip prospers. The population is expanding and businesses are beginning to appear. The name, Colstrip, Montana, is becoming familiar throughout the United States and Canada as each day prominent executives come to examine for themselves the potential of this little town. It is amazing and astounding to see it develop so rapidly before your very eyes, but this is exactly what is happening and after only a short time in the town the idea of it some day becoming the 'Denver of the Northwest' seems very possible and even probable."

"At that point, you know," Barbara remembered, "I wasn't thinking in terms of the consequences."

Barbara Wimer was still in college, when, in 1971, Montana Power Company quietly announced its plans to build a coal-fired power plant at Colstrip. Her thoughts were far from home by then; they were most concerned with a young fellow student from Topeka whose parents ran a dry-cleaning establishment but whose passion was roping calves and riding bulls in the rodeo arena. He had a narrow face, curly hair, blue eyes, and slightly prominent ears. His name was Charlie Wallace.

20

The Montana Power Company announced yesterday that Colstrip will be the site for its 350,000-kilowatt steam generating station. . . . George W. O'Connor, President of Montana Power, said as many as 400 workers will be employed at peak of construction. . . .

The Colstrip site, said George O'Connor, is being designed for possible ultimate power generation of 3,000,000 kilowatts.

—*Forsyth Independent*
July 15 and July 22, 1971

When he went to school in Colstrip, Don Bailey was not known for keeping cool when irritated. "Donny is quick on the temper," a former teacher remembered. "He was the kind of person, Donny was, that if things weren't going his way he was mad, and I mean mad. And now when he gets on television, it's the same thing." In 1971 Don Bailey, who ranches up the Rosebud about six miles from Wally McRae's ranch, decided that the increase in coal mining in Rosebud County was not the way he wanted things to go.

Don Bailey's grandfather, Henry Bailey, came to the Rosebud in 1886, ten years after the battle of the Little Bighorn. One of his pastures contained a flat on which Custer and his troops had camped on their way to the rendezvous with the Sioux and Cheyennes. Eighty-five years later the field had changed very little: There were a few fences and cattle in the foreground, but the land dropped gently down to the same twists in Rosebud Creek, then rose again to the ageless ridges of sandstone and scoria from which Indian lookouts had watched the troops bake bannock and tell stories about victory. But just over the ridges, three or four miles from the old farmhouse, whole formations of rock and earth were already being moved aside in rubble to allow miners to get at the coal. Where prairie and stone had been were now appearing rows of mounded overburden, standing seamed and grass-less next to new-laid railroad tracks.

At first this did not bother Don Bailey—or at least it did not gal-vanize him into opposition. The old Northern Pacific Railroad mines had been similar to this if not as large; the idea of renewed coal mining in the area had seemed to carry such little threat to him as the '60s ended that he and his father had sold nine hundred acres to be mined at fifty dollars an acre. But in 1971 the United States Department of the Interior published a thick book called the *North Central Power Study* (*NCPS*). The book, a major tactical error on the part of the utility in-dustry, took plans that were not much more than speculation, even in some industry boardrooms, and spilled them out across the northern Great Plains and eastern Rockies in a cascade of little pieces of fear. The map of the region published in the book was speckled with these bits, each of which represented a coal-fired steam-powered power plant.

The *NCPS* forecast that by 1980 the area would produce 50,000 megawatts of power generated by coal strip mines near each plant, and by 2000 would generate 200,000 megawatts, power to be sent both east and west to distant load centers on 765-kilovolt power lines. The power, the study said, would be generated by forty-two plants, thirteen of which would produce 10,000 megawatts each.

At the time the *NCPS* was published, in October 1971, the only practical experience Montana had had with steam generation was the Corrette plant in Billings, which produced only 180 megawatts.

"I began to smell a mouse right quick," Don Bailey remembered. "Being as it was put together by thirty-six companies and the govern-ment. It was an alarming production."

The *North Central Power Study* was assembled at the direction of As-sistant Secretary of the Interior James R. Smith. Almost every power company in the Northwest, from Wisconsin to the coast, participated in its various committees. Montana Power was among nineteen inves-tor-owned power companies that participated; it provided three men who served on land reclamation, coordination, and resources commit-tees. The highest-ranking Montana Power officer who participated was W. W. Talbott, vice-president, marketing. In the light of the subsequent negative impact of the *NCPS* on the public, the Montana Power executives no doubt regretted that they could not disassociate themselves from the study.

The people of the area around Colstrip had grown up with coal mining; when the mine closed down in 1958 and most of the men left there was a time of sadness, of breaking up of a tradition of friendship between miners and cattlemen. So when the news originally broke that Montana Power was going to reopen the town, there was little dismay. The old relationship would be reestablished; the children of the two backgrounds would meet and marry as before, and the town would once more be a small but active center of commerce; mine tipples and grain elevators standing side by side. But the power study indicated that instead of being a repeat of the Northern Pacific operation, Colstrip was going to be the new heart of an industrialized West.

For Bailey the *North Central Power Study* disoriented his whole world. It was as if all the intricate structure supporting his life—the friendships, the habits, the places he would go to hunt, the places he went to get away from people, the places he went to party, the political balances he spent youth struggling with and adult life learning to turn to his own needs, the values he leaned on without question to support him and free him from the impossible daily weight of pure freedom, even the very sweep and comfort of the land—were suddenly shown to be nothing more than strips of painted canvas tacked to frames and held up against a void of unknown experience only by pegged guy wires that would soon be severed, one by one, by aliens. Beyond the tumbling facade was nothing familiar at all.

So Don Bailey got mad.

21

MONTANA POWER TO DOUBLE SIZE OF
COLSTRIP STEAM ELECTRIC GENERATING PLANT
Tentative plans are now to install 700,000 kilowatts of power in a two-unit steam electric generating plant at Colstrip . . . requiring investment of more than $140,000,000. [Partner in the venture] is Puget Sound Power and Light Company of Bellevue, Wash. . . .

—*Forsyth Independent*
January 27, 1972

Coal. Anthracite. Bituminous. Subbituminous. Lignite. Hard, soft, seamed, flaky. The tragedy of Wales; the triumph of Newcastle. The waste of West Virginia; the glory of Pittsburgh. Gas for the Luftwaffe; steel from the Ruhr. A mysterious smoky substance once banned in England as poison; the fuel for the industrial revolution. Plastics and black lung. A coal riddle: If you can put 8,700 BTU per ton eastern Montana subbituminous coal on a unit train, how much will it cost a citizen of Cook County, Illinois, to fire twelve clay pots and a gravy boat in a 1,000-degree kiln? And who will complain that it should never be done?

Martin White turned thirty on August 3, 1971. By then that kind of question was not entirely hypothetical in his life; his existence now centered on the extraction of coal from the ground of eastern Montana. What had started as a casual job, a marking of time until he should be drafted, had turned into a career. At first when he had started working for Western Energy in his spare time at the office coal was simply prices and tonnages, but as Western Energy grew White was drawn into it full time and some of the excitement of the company's growth began to replace the gloom that had burdened him since his father's death. The man who showed him which way to aim his reborn enthusiasm was the president of Western Energy, Paul Schmechel.

"I didn't even have an opinion on coal until 1967," White said later. "And from '67 until even '70 my opinion would have been tremendously influenced by Paul Schmechel because I was still just a fledgling young fellow in the company and during that period there wasn't enough controversy about coal for me to think of it too much beyond the fact that people needed it and we were charged with the responsibility of mining it."

Schmechel believed in coal. He believed in it with the fervor of an intellectual theologian, his determination buttressed by his mind. He made no concessions: When others in similar positions in the West apologized for the need for poles to string power across the country, Schmechel called the huge towers "steel soldiers of progress marching across the land." While others bent to the wind of popular rhetoric to talk about the need to cooperate with the environment, Schmechel told an audience that the real wealth in America was produced by "the man who works with his hands and fights nature day in and day out." Martin White was young, emptied by tragedy. Schmechel filled him up. Working for Schmechel, White must have felt like a Burlington Northern railroad car in Colstrip when it rolls under one of the mine tipples and is filled with one hundred tons of coal in ninety seconds.

Although White would never think of himself as a pawn of anyone's persuasion, he began to feel that Schmechel's concepts made sense.

"We'd periodically get in discussions concerning the future of coal and how it should relate to nuclear and oil and gas," White said later. "I'm not sure whether he was trying to convince me or convince himself because sometimes Paul does that—he repeats things and thinks about them and repeats them—he was trying to show why coal really is going to be the number one fuel in the country. And I guess he pretty thoroughly convinced me that it was going to be and he had done enough research and I think as I look at it today he was exactly right."

As the 1960s ended western coal had grown more and more important to the nation's utilities. Although some of the specific projects the group called WEST had promoted were scrapped, and though the *North Central Power Study* was received by such public dismay that the companies who helped write the plan had to renounce it officially, the enormous goal of vast western power remained hanging in the atmosphere like static around conventions of electric generation executives and equipment manufacturers. And various utilities had begun to build mine-mouth or remote generating stations of unprecedented size or to import western coal. Utah Power and Light, the Los Angeles Department of Water and Power, Southern California Edison, Puget Sound Power and Light, Commonwealth Edison of Chicago, Northern States Power and Light, and Washington Water Power were among the utilities involved. The eastern utilities were buying coal to ship home; the western utilities were building power plants. By the early 1970s plants were either complete or under construction at Four Corners; Page, Arizona; Fallon, Utah; the Mojave Desert; Gillette, Wyoming; Huntington, Utah; and Rock Springs, Wyoming. Montana's 180-megawatt plant in Billings had been an early, if modest, entry in the field.

At the same time that utilities all over the western half of the nation were planning new plants fired by western coal, they were also consolidating coal property to feed those plants, in either the broad expanse of the Fort Union formation or in the several basins of coal located to the south in Colorado, New Mexico, Utah, and Arizona. Among the leases taken in Montana were 214,000 acres of land either leased for mining or on which permits were given for prospecting on the Northern Cheyenne Indian Reservation. And just north of the reservation, near Colstrip, Montana Power also expanded its reach. Led by Schmechel and by George O'Connor, Western Energy purchased leases from Northern Pacific and took new leases from the federal government to bring its total reserves to about 830 million tons of coal.

Much of this was being done on speculation of future demand, some of it based on increases in per-capita consumption of electricity. At the time the leases were being purchased, Schmechel said later, "there still wasn't any market for western coal." Montana Power was in the process of building its own market—the Billings plant—but that would only consume six or seven hundred thousand tons per year. But Montana Power had other plans. If other utilities could be persuaded to join Montana Power in the construction of mine-mouth generating plants, the company would have, essentially, a slave market right there at its command.

With the growing interest in western coal and in Western Energy's new properties, it was not surprising that the "short book" that Martin

White had taken on in 1967 turned into a full-time job in early 1968. And as utilities in Chicago, Minnesota, and Wisconsin became interested in the Western Energy reserves it became more than a full-time job. Suddenly it was a career, a way of life. Suddenly he was dealing with power supply all the way to Chicago.

There were sales trips, long telephone calls, and negotiations. Northern States Power Company was interested in converting that switch, wasn't sure how well the 8,700 BTU coal would burn in furnaces originally designed to handle more powerful Illinois bituminous. Northern States bought 10,000 tons in 1969, 480,000 in 1970, and just 920 in 1971. Commonwealth Edison in Chicago was even more skeptical. "Burning your coal," one executive told White, "is about like burning leaves." But Commonwealth Edison bought 730,000 tons in 1970 and 3,700,000 in 1971. There used to be a small underground coal mine near Billings called the Smokeless and Sootless mine; after comments like Commonwealth Edison's, someone at Montana Power, perhaps jealous of the upstart subsidiary, suggested they ought to call the Western Energy operation the Smokeless, Sootless, and Heatless mine.

There were occasional failures: after courting Minnesota Power and Light Company of Duluth, for instance, with promises of the Rosebud seam's capabilities, Western Energy was underbid by Peabody Coal, which mines the same seam just a couple of miles down the Lame Deer road. But win or lose the work was demanding. For several years Schmechel and White were the only employees of Western Energy—the mine work was contracted out—and, White said, "I can remember many times when Paul and I would be in the office until ten o'clock at night."

If White's memories of his father, who was afraid all the land would be consumed by industry, irritated the back of his mind at all, they were offset by the grandeur of the drive toward western coal. The coal brought him stature and financial security; but more: It drew him into the national arena whose edge he had stood on during the Olympic speedskating trials. He had fallen into a chance of a lifetime: It was as if he had reached out in curiosity, like a caveman, for a strange, seamed lump of black in an unassuming crevice and found it shot with flame, lit and blown to white by the sudden demands of two hundred million people. And the blaze was in his hands, the heat of the desire on his face, and the power in his body to help throw the fire forward.

As young men elsewhere in the United States—including Vic

Jungers—found themselves caught up in the drama of being a part of the national political upheaval of the antiwar and civil rights movements, so Martin White was captured by the nation's relentless drive for energy. Suddenly he was tied to an electric fire siren in Chicago, a dentist's drill in Duluth, and a shopping center in Billings. He was a man of power.

And in 1972 the Western Energy mine delivered 5,500,700 tons of coal to its customers and was officially recognized by the U.S. Bureau of Mines as the third largest coal mine in the United States. But White and Schmechel didn't celebrate. "We were too busy," White remembered. "We didn't even have time to stop and drink a beer to it or sit back and swell our chest and say 'Gees, we did a hell of a job.' When they said we were the third largest mine in America, I thought 'Holy Jesus!' and went back to work."

It was a long way from The Dry. It was a long way from the hills above Butte where the dead elk had steamed and the noise of the distant jeep had warned White's father that the machines of men were reaching out to conquer the world. Today White stood on the inner curve of the racing wheel of industry, and the acceleration leaped in his heart. Like Schmechel he became an apostle of coal. "By 1972," he remembered, "I was convinced that coal was probably the only fuel that we would be able to meet the electric generation requirements of this country with." Harrington had been right; White had spent his sadness in his work, and now the desperation was gone, swept away in these new exhilarations.

But Martin White was no longer the cheerfully critical youngster whose attachment to the company was only temporary.

"Martin White's loyalty transcends every other quality the guy has," Paul Schmechel said. "He's got a super personality; he presents himself well; people like him and they trust him. But beyond all of those great things, his loyalty is just fierce."

So by 1972 Martin White had considered the evidence and had made up his mind.

22

George O'Connor, president of Montana Power Company, predicted coal would become a giant in Montana's economic future as he signaled the start of construction on the utility's new 700,000-kilowatt plant.

—*Forsyth Independent*
May 18, 1972

The Student Union Building at Montana State University in Bozeman, better known as the SUB, was awash in people. The air held the sparkle common to places where famous individuals rub against the people. Today the guest speaker was a man named Russell Means, a leader in the American Indian Movement.

Marie Brady Sanchez, Wandering Woman, was both a student and a teacher at the university. She taught a course in Northern Cheyenne language and culture. But it was more than scholarly interest that made her join the crowd waiting for Means. She longed to meet him, to find in his eyes some solution to her own dilemmas about the Indian role. Certainly this man held some answer, could offer some direction, some hope. "I was trying to at least get a statement that would stick clearly in my head for the rest of my life," she remembered. "That would *determine* my life." But there was no knot of people yet in the room to show where the great man was.

By 1972 the Northern Cheyennes had leased out about half their reservation to several coal-mining corporations. The price had been low and the negotiations conducted by the Bureau of Indian Affairs had been minimal. The resource was there and would be extracted, and the Indians would receive a skim off the top of the profits, little more. It was a traditional arrangement.

But there were changes growing among Native Americans across the nation, and they were bound to be reflected in the Northern Chey-

85

enne tribe before long. Marie Sanchez' presence here in the SUB waiting for the word from Russell Means was an indication that the time
was near.

She wandered vaguely through the crowd. Then she saw him. The
classic Indian profile. The hair pulled back in twin braids. The flashing dark eyes. Stature, poise, wisdom. She worked her way through,
and when the man stopped talking to someone else she interrupted.
"Excuse me, are you Russell Means?"

The Indian laughed. "No," he said. She was embarrassed. "Here,"
he added, "I'll introduce you." Through the crowd a newspaper was
approaching, held high before an invisible person, like the prow of a
ship in the crowd, splitting groups.

"Hey, Russell," the Indian said to the newspaper. "This is Marie
Sanchez, Northern Cheyenne." She listened to her name. It was like
the sound of the crier on the field at Busby. Marie Sanchez, Northern
Cheyenne, Wandering Woman. Suddenly her race, her relationship
with these men, had meaning. A whole meaning, for a race. The river
of new movement that they stood in washed around her knees. She felt
its size, its grip on all Indians, its coldness, and its demands, as it swept
at the heels and thighs and bodies of the Navajos, the Crees, the Sioux,
the Chippewa, the Blackfeet, the Kiowa, the Crow, wherever they were
across the nation.

The blood in her stopped. There was pressure in her ears. Her life
waited, like pent flame held in the bottle, longing for the hand on the
valve and the match.

The newspaper lowered, slowly.

Russell Means looked at her.

She waited.

He spoke.

He said: "Oh, hi. Glad to meet you."

He went back to his newspaper.

The other Indian disappeared with him. Sanchez was left standing in the crowd. It flowed around her, ignoring her. She watched the
newspaper disappear. The great man. She had caught a glance of the
article on the page that he was reading. He had been reading about
himself. She smiled very slowly, very softly, so no one around her could
see the triumph in her eyes.

"At that point when I met Russell Means," she remembered
much later, "I felt—I felt: Now *look*—for once here's an Indian who
isn't on his knees."

23

If I were the owner of land in Southern Rosebud County, and sitting on top of a vein of coal, not having the facilities to mine the coal, and if I could make more money from selling my land to a coal company than by grazing cattle on it, I would sell.

—Editorial, *Forsyth Independent*
June 29, 1972

Looking back on the start of the conflict over coal and power, it seemed to Wally McRae that in the beginning it was not much more than a game. On one side were the coal men, usually Paul Schmechel and George O'Connor, promoting economic growth and trying to emphasize the temporary nature and limited size of the holes they were digging in the countryside; and on the other side were the cow men, Wally McRae and Don Bailey, usually, promoting the temporary nature and boom-and-bust economy of strip mining and trying to emphasize the value of unstripped land to the cattle industry and to the essential meaning of western life. They would go to the meetings that were popping up all over Montana like pimples on an unquiet skin, they would line up at the start, rage at each other for two hours by the clock, then shake hands, grin, and depart to buy each other dinner.

This curious blend of public animosity and private friendship was a kind of traveling road show that played more and more engagements as coal grew as an issue in Montana. McRae spoke at a coal symposium in Billings, at a governor's conference on mined-land reclamation and Montana mining law in Helena, on television panels, and at Chamber of Commerce meetings. Often he'd travel to and from the meetings with another coal spokesman, Gene Tuma, of Forsyth, who worked for Peabody Coal.

87

"Gene'd call me up or I'd call him up and say 'Going to Glendive, going to Billings, going to Miles City—do you want to drive together or do you want to drive separate?" McRae remembered. "So we'd fly together or drive together, just to share the load. And we'd visit all the way there and visit all the way back and scream and fight at one another in between."

Once he showed up at a meeting in Bozeman in the middle of a winter storm and Schmechel, one of the panelists, was late. The moderator waited, and waited, and finally Schmechel walked in, having fought the road down from Butte. McRae was the only individual Schmechel recognized in the group, so he gravitated to him, shook his hand, and complained about the road. "Three hours to go a hundred miles. Didn't think I'd make it."

"That's all right, Paul," McRae said. "If you hadn't showed up I'd have given my speech, and then I'd have given yours."

The two men laughed. Five minutes later they plunged into the battle.

McRae hated to make the same speech every time, no matter how far apart the meetings were held, so he would write a new one for every occasion. He would work late into the night, honing fierce statements of heritage and anger until two in the morning, then reading them the next day at high speed, like a man firing a machine gun continuously until the barrel melted. But each speech would, of necessity, contain familiar elements: the beauty and value of his home country contrasted to the degradation brought by strip mining.

"There is a myth about my part of the state," he often said, "that has it fixed in the minds of many people as a semidesert, with endless flats, barren, and treeless; broken only by an occasional badland gumbo butte. The impression is that this is a country so devoid of beauty and productivity that it is only fit for cactus, cowboys, and catfish. And all three of these rather undesirable species wish they could live somewhere else.

"In reality this is a beautiful, productive land, with an abundance of timber, live streams . . . and unsurpassed rugged western beauty. The desert myth cannot stand up to the fact that there was a timber sale in Rosebud County this past year for five million board feet of timber. . . .

"Strip mining as practiced in the past has completely destroyed the productivity of the land for agricultural crop and livestock purposes, it has also destroyed the land's productivity for wildlife and game, and has perhaps even more significantly, and importantly, destroyed the natural beauty of the land."

"It *was* kind of a game for a while that we played," McRae remembered. "And it was a serious game. But I felt that my point of view should darn sure get out. And I also felt that their point of view should get out . . . because a lot of times their point of view offended people. And that suited me like hell."

Part of the game, perhaps, was the expression of his cause in his poetry.

Millions, or billions of dollars, or tons,
It just doesn't matter to me.
I'm too old, or too young, too rich, or too poor,
Too smart, or too stupid to see . . .

They're turning my country all wrong-side up,
In the quest for their black gold . . . coal.
Ravaging the land just wrenches my heart.
A million acres, their goal.

Why, if this is our country's salvation,
Do I recoil with disgust?
Why does the flavor of gold in the air,
Taste so, of ashes and dust?

I should rejoice with each earth-shaking blast.
The coal-smoke incense should thrill.
What difference the spoil pile, conically shaped,
And the cone-shaped red scoria hill?

Is it the memory of forebearers dead?
The question in my son's eye?
Is it just pride, that won't let me give in?
Why, dear God tell me why.

Millions, or billions of dollars, or tons,
It just doesn't matter to me.
I'm too old, or too young, too rich, or too poor,
Too smart, or too stupid to see . . .

Then one day McRae drove past Colstrip on his way up to Forsyth, and there down by Armell's Creek in a little basin between a ridge of old spoils and the town itself, was a wide concrete foundation;

huge and menacing, like a sphinx or a pyramid suddenly appearing on the desert. It covered acres. It was swarmed over by trucks and people. It was growing, shaped by plywood and stiffened by a mesh of steel. He had known this was coming, but its physical presence had at last reached the point at which it dominated the town and all the land within sight. This was something new, too big anymore for games. McRae drove past, feeling suddenly less powerful, less confident, less at home in his own county than he had felt since the day he was born. In past meetings and interviews he had often talked of the "inexorable power of coal"; it was a good rhetorical device, painting an image of the embattled cowboy fighting the inhuman power, and he had never really had to believe it. But not long after he saw the concrete caissons planted in the valley beside the old town the phrase began to disappear from his oratory. Perhaps he tired of it, thought it counterproductive; or perhaps it began to sound too true.

24

Ranchers have charged that Montana Power is dodging the law while developing political muscle in the construction of a mine-mouth generating plant.

An insistent crowd of ranchers, ecologists and local residents grilled the state's air pollution control chief, Ben Wake, demanding action. . . . Montana Power Company has been at work for several months, constructing a giant mine-mouth steam plant. . . . To date, the company has not applied for a permit. . . .

Wake admitted the company's failure to obtain the required permit, but said he had assurances from Montana Power officials that the new plant would meet and exceed Montana air pollution control standards.

Wake was asked by rancher Don Bailey how he knew they would do the things they promised to do.

Wake responded: "I don't know why they [Montana Power officials] would lie to me."

"If the company's . . . structure doesn't meet state clean air standards, do you think your agency will have the political muscle to tear it down?" Bailey asked.

Wake said he didn't think it would come to that.

He said the application for the permit would come within two weeks.

Wake argued that by not requiring the company to obtain a construction permit he was allowing the company to take advantage of developing technology— technology that will result in a cleaner operation when the plant begins to turn Eastern Montana coal into electric power.

The plant is the first of seven planned for Colstrip.

—Forsyth Independent
August 3, 1972

The courtroom was not crowded. There were less than thirty people in the room. They made little noise. The radiators squeaked, and there

91

was a murmur outside from the county commissioner's office across the hall. It was December 1972.

If there was any place in time in which the conflict over coal stopped seeming like a game, it may have been here.

Across the state the game had gone on: television debates, panel discussions, and symposiums had grappled with the issue of coal development in Montana, conveniently focusing on the place where conflict was most well defined, Colstrip. But today, when substance itself was at issue, most of the players in the game of rhetoric were absent. Wally McRae, who was one of the few members of the Colstrip Traveling Road Show present, later called it a non-event. And yet it was a turning point, a place where the conflict was joined on a new level. For the first time the plant, before it was much above the surface of the ground, had brought the matter to court.

Soon after McRae had seen the plant rising above the surface of the ground like a huge stone whale, he had joined an organization called the Northern Plains Resource Council (NPRC).

The NPRC was a curious organization. Just a few years before, when the Montana Stockgrowers Association was passing resolutions condemning student activism on college campuses, such an organization, a loose coalition of ranchers and young environmentalists, would have been inconceivable. The ranchers had been galvanized into action by their anger at the coal companies' intrusion into their domain, and the environmentalists were refugees from the struggles of the 1960s. The ranchers were the official membership, the political front, and the leadership, and the environmentalists were the staff; once a month they met in Billings in a tiny office across the street from Montana Power's imposing Billings office, to plan strategy.

McRae had decided to join the NPRC not long before the suit was filed. It had not been an easy decision.

"I was announcing the rodeo at a fair and they had a booth and I talked to them and they said, 'Why don't you join?' Boy, I didn't want to join that outfit, that bunch of wild-eyed, fuzzy-headed environmentalists. I said, 'Man, I don't know. I think that I can do more good as an independent rancher, talking to other independent ranchers about the threats of coal development and what we've got to do.'"

But McRae realized that he would probably never get anywhere just by running surveyors and core drillers off his land. His own ranch might be spared, but the society of the Rosebud that he valued could be destroyed. "I could build my own little island here if I wanted to," he told one of the reporters who were starting to find their way down to the Rocker Six in search of a photogenic face and a sharp-edged quote. "But I'm not sure I want to." The quote may have been one more concession to the image of the lonely cowboy that McRae knew had propaganda value but limited real power; he had no intention at all of becoming an island. So he joined the NPRC.

"I finally kind of decided I'd be a closet member and wouldn't brag about it. And about that time they were building this dang thing over there."

But McRae was not good at hiding in closets. His eloquence got caught in the door. So the closet membership in NPRC had bloomed into something close to leadership of the organization by the time the NPRC took on the power plant.

The first battle of what was to become a protracted legal war was fought, as would be the last, over a permit. Typically, this matter, which both sides thought of as a moral issue involving major philosophical attitudes toward resources, actually reached confrontation over a narrow point of law. The law in this case was a requirement that any individual or corporation had to get a permit from the state Department of Health before building a polluting structure. McRae and the NPRC thought this would naturally apply to the plant, but they soon found out that no such permit had even been applied for.

This was in the last days of corporate control of the state of Montana. Anaconda Company was still chewing the heart out of Butte, and industry still thought it was the most powerful force in the state, still invested with the special privilege that had allowed it in the past to ignore laws that were inconvenient and create those that greased its wheels. So when the NPRC demanded that Montana Power get a permit from the Department of Health, the company was mildly amused.

A permit from the Department of Health? Well, the company said, we're not building a polluting structure—we're just pouring foundations. We'll get the permit in good time—about the time we're ready to install a boiler. *That's* the potentially polluting structure, not these concrete footings. The argument was a vital one, and would have curious reflections later, black on white: We have no need for a permit because though we have begun construction of something we have not yet begun construction of the polluting device.

At the time McRae still had some faith that the gears of government were lubricated by reason. "I guess I was naive," he said. "I was appalled not only that the company would violate a very obvious law, but I was even more amazed that the state Department of Health itself and the judicial system within the state of Montana wouldn't enforce a very plain specific piece of legislation. . . . And that it took a citizen's group to get the department and the judicial system and perhaps the largest corporate entity in Montana to even pay any attention at all to this law. I was flabbergasted."

McRae's first chance to express his frustrations came at one of those informal panel discussions on coal and pollution that he had been attending all over the state. Only this time he was not a speaker. It had been held in Helena, in a grade school auditorium. McRae had waited, restive, in the audience, until the state's director of air pollution control, a panel member, was available for questioning. Then McRae stood up, dark and slender in the crowd. This is the way he remembered it:

"Do you feel that the 1971 Air Quality Act is being enforced consistently?" he asked.

The director said yes.

"Okay," McRae said, employing the tone of weary patience he sometimes assumed, as if he was trying to lead a class of first graders through the principles of addition. "Okay, now your department has found that Montana Power is constructing what I feel is a potentially polluting facility at Colstrip. They don't have a permit and your department has told us that they don't need a permit until they either install the devices that have the potential of polluting or possibly later on in the construction program than that, when they install the pollution *abatement* equipment. And I understand there's also a possibility that you might extend this until they start *burning* coal which has the potential of polluting the environment."

The director said, "That's right. We have no ability to turn them down and declare that a potentially polluting device until we find out how much it's going to pollute, what kind of boiler they've got, what kind of pollution abatement equipment."

"All right," McRae said. "Let's look at it another way: I don't think there's any doubt in anybody's mind but what a cattle feedlot on a riverbank is a potentially polluting facility."

The director said he would agree.

"All right. I'm a livestock person. There's a river running by my place. If I want to build a feedlot on that riverbank, now when do I have to get a permit? If you hold the same rules all the way through, I don't need a permit to build the facility until I install the device that has the potential of polluting the environment—which is *the cattle.* Okay, now my question is: When do I have to have a permit to build a feedlot on the bank of a river?"

The director said, "Just before you dig the first post hole."

"Thank you very much," McRae said, and sat down.

But the confrontation was just a rhetorical exercise. Nothing was done. So, finally, the NPRC filed suit against Montana Power. After the company asked successfully that the trial be moved from its original venue in Yellowstone County to Forsyth, it came to trial in December. Both McRae and Don Bailey testified, but McRae remembered little of the circumstances except the behavior of the Montana Power lawyer, which stuck with him through later years as an example of the general stance of industry.

"His attitude was one of utter disdain," McRae remembered, "that this was such an exercise in futility for him to have to get in his car and drive down the road to Forsyth, Montana, to pursue something that was so blatantly ridiculous. And every time there would be any kind of a thing at all he would very slowly get to his feet and shine his bald head with his hand and say: 'Your Honor, I move to dismiss these hearings.'

"And Judge Coate would say: 'I will take your objection under advisement but the proceedings will continue.' "

Before the final judgment was issued several things happened. The state Department of Health issued a perfunctory environmental impact statement on Colstrip Units One and Two; Montana Power requested an appropriation of fifty-nine billion gallons of Yellowstone river water a year—enough for at least thirteen more plants the size of the one under construction; and the *Billings Gazette,* which had so praised Montana Power's efforts in developing coal in 1968, called the two units the beginning of the destruction of eastern Montana. If the plants were built, the newspaper said, they "will unquestionably multiply. One monster plant will beget another and another at Colstrip." The point was already well understood by both sides down in Rosebud County: On March 7, 1972, thirty-three landowners had met with three Montana Power executives and had been reminded again that Units Three and Four, of 700 megawatts each, were possible, and that the ultimate possible size of the Colstrip plant was 3,000 megawatts.

There was the fear, concentrated. The threat blossomed over eastern Montana in steam and smoke. From those foundations growing up from the ground in Colstrip what vast creature would arise, attended by a flotsam of boomers, junkies, labor union bosses, discontent, and degradation? McRae was still arguing with some of his neighbors who accused him of turning liberal, of going over to the environmentalists. "But it is not the environmental things that concern me the most," he told them. "It's the social things. It's the massive industrialization. It's just that the environmental numbers are all we have to fight them with." It was the idea that here on this cherished land might arise a place where the dominant force was not man or wind or animal, but machine.

Judge Coate came to one decision in December. If he was to issue a temporary injunction against Montana Power and Puget Sound Power and Light, he told the NPRC, no doubt the companies would lose a considerable sum of money. Therefore, the judge said, if he granted an injunction, would the plaintiff be prepared to supply a bond for the amount Montana Power said it would lose by delaying construction one year? The amount was $44,963,000.

"And we said 'Er, what?' " McRae remembered. "I couldn't believe it. I thought, you know, is this the way our justice system *works*? There was no doubt in my mind whatsoever that we were right."

The demand for the forty-four million dollars was essentially the end of the event, though it dragged on. Judge Coate eventually dismissed the case, but not before a Montana Power ad promoting the Colstrip units, illustrated by a hand holding a match, asked readers of the *Forsyth Independent*, "Will 1975 be the year of Montana's first brownout?" Coate announced his decision in January 1973, saying that the Northern Plains Resource Council had "sustained no wrong, injury or damage from the construction work complained of; and no such wrong, injury or damage is imminent . . ." and that no air pollution had resulted from the construction work currently underway. While the power companies had ordered materials from the manufacturers that had a potential for causing air pollution, he ruled, the articles had not been delivered and were not being installed.

The principle that construction of a polluting structure had not begun until the actual polluting instruments were in place had been ratified in court. But there would come a time when, in another court,

Montana Power would choose to argue another way. And the power plant would win again.

In 1973 the state legislature stiffened the rules for the construction of new power plants. Helped along by the lobbying efforts of McRae and the NPRC, it passed the Utility Siting Act, an act that required approval of proposed facilities by the state Board of Natural Resources and Conservation after it determined that the project was necessary, that the facility represented the minimum possible adverse environmental impact, that the facility would serve the public "interest, convenience, and necessity," and that it would not violate state and federal air and water quality standards. It was a victory for those who wanted more controls on power plant development in eastern Montana, but the triumph was qualified:

One provision that McRae and others had proposed was a three-year moratorium on new construction. Although Colstrip One and Two would not be affected by the moratorium, Montana Power fought the proposal with all its resources. The moratorium was defeated.

25

For several days now Duke McRae, riding out to check his heifers in their summer pasture, had seen the tracks down in the little canyon. He was a long way from the house, which lay down on the Rosebud in a stand of silver leaf maples, but up here he was not far from the road to Colstrip, and the tracks bothered him. They were narrow tracks, the kind made by a car rather than a pickup, and they wandered up across the soft soil of the draw and stopped in the trees near the top. The first day he had noticed them he hadn't thought too much of the intrusion onto his land; with the plant going up in Colstrip there were always a few wanderers who strayed up onto private land, and Duke wasn't going to be the one to shoot a stranger for an honest mistake.

But the next day when he looked at the tracks they were overlapped by another pair. McRae stopped his pickup and got out. They were the same kind: narrow tires with a road tread that looked feminine beside the knotty grip his truck tires left in the earth. He walked up the little slope to where they stopped in a grove of small Ponderosa pines. There was the back-and-fill story of a vehicle that had turned around, and there were also footprints. In contrast to the tire marks, these tracks showed authoritative cross treads, but the oddest thing about them was that they disappeared on up the draw and did not seem to return. And each footprint had the characteristic emphasis on the toe which shows the person is running.

McRae shook his head and walked back to the pickup. And for a week he drove past the spot every day, varying his times, morning and afternoon, and all he found were more tracks. And each time he looked there was another set of the same kind of footprints running out away from the grove and never seeming to return. There was never any trash in the grove, none of the beer cans or marijuana cigarette butts McRae had sometimes found where the coal company surveyors had spent just a couple of hours, but it was clear to him by the second week that someone odd was living here on his land.

While the mines had expanded around Colstrip, the trenches reaching out and turning over land that had once been his, and while the power plant had been growing on the north side of his ranch, Duke McRae had taken a passive role. He had made no public stand, although he knew his sympathies lay with the cause his cousin, Wally, was turning into a crusade. But the little things that had goaded Wally into action—the arrogant carelessness of the core drillers, the gates they left open, the dope they dropped in his fields, the lies some of the coal surveyors told to get permission to go on his land—only made Duke McRae sad, not angry. He was more worried about the subtle impacts of the things going on over the hill, knowing the roundabout ways the earth has of punishing mistreatment, but even these took a long time to make him decide. When the renewed interest in coal had promised a rejuvenated Colstrip back in the '60s he had even sold some of the ranch to be mined; he was not going to turn around and fight that decision until there was no choice.

"The thing is," he said, "if you could go out there tomorrow and you've got fifty dead heifers out there, and you could lay it onto that plant, by God you'd have something to go on. But it's not going to be that way. It's going to be a slow, gradual something that you can't ever, really, I don't think, completely prove."

Like the water. The coal seams in the Fort Union formation are often the best aquifers—the porous rocks that carry the water along beneath the surface. What happens to the water levels in springs and wells in the area of strip mines is still unknown. It is certain that the aquifer is affected; in the bottoms of the mined holes ponds collect in a country in which lakes are unnatural; they look like pools of plasma lying stagnant under the cliffs of spoil.

A Montana hydrologist studied water levels near a strip mine to the south of Colstrip in the early '70s; the study indicated that, as a result of mining, wells are likely to drop up to several miles from the trench. In an impact statement written later, the same hydrologist is quoted reporting that the increased mining needed to feed a 1,400-megawatt power plant at Colstrip would destroy as many as nineteen wells and three springs in the mine site, could severely lower the water in four adjacent wells, and could significantly lower the water levels in five more nearby wells and two springs. As a result of the mining, the impact statement said, "The groundwater system, since both coal seams and the sandstone layers in the overburden are the major aquifers in the area, will be disrupted indefinitely. If the groundwater flow establishes a new pattern the quality of the water may change due to flow through different strata."

To Duke McRae, downstream on the aquifer, the effect of the mines on this blood of his livelihood might be the most significant of all, but the slow drying away of the water may be as hard to spot as the first gray tickles of radiation disease in the brain. If the life is sucked from the ground over the years, or tarnished by new rust, the relationship between the cause and the effect would be almost impossible to prove: Only the damage would be sure. So you watch your windmills pump the silver from the earth while they can, and you watch the cattle drink, and you wonder when the day will come when all that will come up from the ground will be sand.

One night about two weeks after he had first seen the tracks in the little draw, McRae went out in the cool dusk. The air along the creek was fragrant with the smell of the hay growing lush in the moisture of the creek-bottom soil. The hills had the pale softness that they wear after the sun has taken its furrowing hard light down over the horizon. Summer was advancing; already the green of the pastures was fading out into a wash of yellow-brown. A meadowlark somewhere in the air broke the silence with its strangely melodic, intricate song, a tune that always seemed too rich for this simple land. A cat was sitting on a pole in the corral like an owl. McRae got in his pickup and drove it up out of the bottomland, stopping several times to open pole and wire gates, drive the pickup through, and close them carefully behind him. The cattle he could see in the vast pastures were not moving; they stood or lay quietly, absorbing the cooling air after the assault of a summer day.

The pickup thumped and rumbled up over the low hills, and a thin mist of dust rose behind it and settled slowly back down in the place from which it came. And when he crossed the last little divide, there, down in the grove where the tracks had been, he could see the shape of a vehicle, half hidden by the trees. It was just dark enough so he could tell there was a light on inside.

McRae drove slowly down into the draw and stopped. He climbed down from the truck, a stocky figure in jeans and a long-sleeved shirt loose at the cuffs. He slammed the door of the truck. The noise was brief and crisp in the still air. The vehicle in the trees was a camper van, with its top pushed up on a tilt like a can of packed meat that has been pried open with a large knife. On his side there was a louvered window, slightly open, but all he could see inside was light.

McRae walked around the vehicle. The light from the open sliding door spread a rectangular glow on the grass, and McRae realized it was getting dark. He walked to the door. There was no sense in being sly. He just walked up and looked in. There was nobody there.

The van was well equipped. The electric light in the ceiling illuminated a Formica-topped table, a small refrigerator, a stainless-steel sink, a stereo tape deck suspended from the ceiling, and a sleeping bag rolled upon a foam mattress in the back. There was a little closet with a mirror on the door, and stuck in the side of the mirror was a postcard photograph of cliff dwellings in Arizona.

There was a sound behind McRae, the thump of feet and heavy breathing. A voice said, "Hey!" McRae turned. A man was standing behind him. He wore a blue uniform made of a soft, baggy material, which had white stripes down the outsides of the legs, he wore small striped shoes, his hair was disheveled, and he was sweating. Duke looked him over. The man said, "Hey," again, then: "What are you doing to my van?"

If Duke had been Wally he might have given the trespasser fifteen minutes to get off the outfit. If he had been another of the ranchers down the Rosebud he might have shot out the tires of the van and made the man run. But instead he took a cup of coffee that was offered and listened to the man explain. He was a welder from down at the power plant, he said, and he hated the confinement of the town. So every night he'd drive up here and camp; then in the morning he would pull down the pop-top and drive back to work. "I close the gates," he said, "and I don't leave any trash. I go jogging every night up the draw here and around that hill and back through the trees over there, and I always run clear of the cattle. I figured you'd find out sooner or later, but I didn't figure I was doing you any harm."

Duke listened while the man, whose name was Fritz, went on. He told him of a life of wandering, of owning motorcycle shops in several different cities, of studying geology, of welding steel at the power plants that were going up across the West. To McRae it must have seemed an entirely foreign kind of life, fascinating as well as repelling to a man who had spent almost all his life on the many acres of the Greenleaf Land and Livestock Company. Finally McRae was won.

"He was a kind of an interesting, nice kind of guy," he remembered. "So I just told him, I said, 'Okay, now, as long as you're here I'll just let you stay here. You watch this for me and just chase everybody else out.' "

So for a little while the rancher and the welder were friends. Fritz took to leaving his van in the draw and riding his motorcycle to work, and often he would come over to the ranch and drink coffee, pet the cats, and talk about how pretty they were. And McRae, working his cows, would sometimes see him jogging in the evening light, up over the little ridges and past the scoria bluffs. So for a short time Duke had contact with the thought and muscle of the construction of the power plant, and it pleased him. There was something less threatening in a structure that thus became human.

But Fritz was a boomer; the life that fascinated McRae would not let the welder stay, and one day in winter he heard of a job in Hawaii. Suddenly the draw was empty, the tracks of his running were swept under the snow and not renewed. It was not long after that that Duke McRae, like many of the other ranchers in the country underlaid by the Fort Union formation, decided he had to make up his mind.

26

On January 23, 1973, executive vice-president of Montana Power Company, Joseph McElwain, attended a meeting of the Montana Board of Health and Environmental Sciences. During the meeting he was asked a question about his company's plans for the future: "Are you considering building Colstrip Units Three and Four?"

He answered: "That is not true."

Three weeks later George O'Connor announced to the Montana Power Company board of directors that the company and three other utilities were studying the possibility of building two 700-megawatt plants at Colstrip. The Northern Plains Resource Council later learned that a joint letter of intent for the construction of Colstrip Units Three and Four had been signed by Montana Power Company, Puget Sound Power and Light Company, Portland General Electric Company, and Washington Water Power Company on September 1, 1972.

In June 1973 the four companies that had signed the letter of intent plus Pacific Power and Light Company filed formal applications with the state of Montana for permits under the new Utility Siting Act, for a 1,400-megawatt coal-fired electric generating plant to be built at Colstrip.

"Based on in-depth studies we have already conducted," O'Connor said in his announcement, "we are utterly convinced the Colstrip plants can be built and operated with minimal environmental effect." He added that the proposed plant should be fully operational by July 1978 or "we in Montana and throughout the Northwest face the bleak prospect of power deficits that translate directly into lost jobs and productivity over a wide range of economic activity."

O'Connor's latter remarks were addressed to the Utility Siting Act's requirement for a six-hundred-day study period between application and the start of construction, a time during which all studies and the issuance of permits would be completed. The consortium requested a waiver of the six hundred days. Montana's governor, Thomas Judge, would not guarantee such a waiver, even though one Puget Sound Power and Light vice-president applied the pressure by saying in March that a six-hundred-day delay in starting the project would kill it.

No doubt none of the officers of the five companies expected that they would ever reach the point at which they would have to eat those words.

27

Residents of old, half-dead mining towns in the West tend to hear any promise of new boom through filters of skepticism. It was that way in and around Colstrip even after the first coal was shipped and the construction of the power plant started. The old attitude died slowly.

"I can remember we talked about it: 'Ha, ha, there are going to be three thousand people here in five years,'" Patricia McRae remembered. "And it seemed so preposterous no one could take it seriously. And then there were three thousand pople and then it wasn't a joke anymore and you could see how easily it could happen."

So at last Duke McRae was forced into action. He would have been false to his family name if he had not spent time in contemplation before he made his decision, but now it was clear that in spite of his distaste for politics, he was going to have to get involved.

"I sat down and tried to analyze myself, which way I thought a person should go," he remembered. "The argument was whether it was going to get that big. I think that probably when we really got involved in it, it had got to the size where they had made application for Three and Four."

In 1973 Duke McRae joined a group called the Rosebud Protective Association, which in turn became an affiliate of the Northern Plains Resource Council. And while he was making up his mind the same thing was happening all over the coal lands of Montana as ranchers began to believe that this thing was the new assault upon their way of life, and this time they, like the Indians before them, were threatened with a fate that smelled like extinction.

For the ranchers the step into political organization to fight a specific issue was fundamentally new.

"One thing people don't appreciate," a staff member of the NPRC said once, "is that the ranchers have for a hundred years been staunch, arch individualists. Oh, they'd get together at the market,

105

when they sold cows, but they never really worked together. And now they've realized that politically and every way they don't stand a chance if they don't get together."

The application for Colstrip Units Three and Four finally got them together, and through them the Northern Plains Resource Council spread a thin but durable net of organized opposition across the country of the Fort Union formation. By the summer of 1973 the NPRC's affiliate memberships included eight other grass-roots organizations like the Rosebud Protective Association. These groups gave the NPRC the kind of breadth given to a river by its many tributaries: It could not be accused of holding narrow interests. And the diversity of the affiliates gave the joint organization an unusual stature among groups broadly labeled environmental: The bond between the young, often liberal staff members and the older, conservative ranchers was tempered by the ideological sacrifices made by the individuals of each side—the ranchers associating themselves publicly with long-haired kids who had participated in the turmoil of the '60s, and the staff members looking for leadership to people who had voted for Goldwater and Nixon and who still shot coyotes—and the effect was a hardening of the mettle of the organization. Liberals in other parts of the country who noticed this odd alliance assumed that the old conservative cowboys had finally turned liberal, radicalized by coal, but the situation was hardly that one-sided.

"They've done as much of a job on our ideas as we have on theirs," said Bill Mitchell, a Northern Plains staff member and a graduate of the University of California at Berkeley. "Hell, I used to be a wild-eyed environmentalist, and being educated in the field I could rationalize my position fairly scientifically. But, like, I've got some fairly conservative views now. . . . These ranchers are basic, conservative capitalists. I respect the fundamentalism of their capitalism."

So from this union came a quiet kind of durability: individualists linked, political opposites allied. To most of the people who favored the power plant the NPRC and its affiliates must have seemed a feeble show of resistance in 1973. Paul Schmechel, president of Western Energy, saw the threat, but perhaps he underestimated it by pigeonholing it too neatly. "You saw a cloud then," he remembered, "and that cloud was environmentalism." But it was not really environmentalism in that word's political sense: an extremism of nature lovers. It was something else. It would take a direct confrontation with the power plant to release its full and varied strength.

28

"But there are too many things that are still changing in the world," Sheila McRae said. "Things are evolving right now, they didn't just *start* evolving. Sure, it had to be created, something was created, but it evolved into what it is now."

The teacher's voice was slightly amused. "Do you really feel that your ancestors were once apes, Sheila?" There was a titter in the room. Sheila was the daughter of Duke McRae, and among the newcomers it was already well known that the McRae clan was something special in Rosebud County. The teacher continued: "What sorts of things are still evolving?"

"Oh," Sheila said. "Cows. You're always breeding for better cows. And it works. A hundred years ago the cows didn't used to be as fat as they are now."

"Neither did Bernie." The voice came from the back of the room. There was a brief scuffle, a large boy turned in his chair and made a familiar gesture, and then there was peace again.

"And if we were apes a million years ago, Sheila," said the teacher, "what are we going to be in another million years?"

The same voice muttered in the back: "Dead."

Sheila McRae turned her slender face to the side, and showed the back row a haughty profile. Her voice was prim. "I *hope*," she said, "that it's something better than we are today."

While the arguments between evolutionists and creationists raged in the Colstrip High School biology class (Sheila thought the teacher favored the creation theory, but she wasn't sure), a half mile east of the old brick school building the power plant inched upward, silhouetted against the spoil piles and scoria ridges on the horizon. Each day when school was let out and the growing torrent of kids poured back into the community there were a few more children to go home and a few more

107

feet on the plant, but neither could be seen at each day's passage. But a week, or two weeks, showed the difference. The scaffolding was a level higher, there was more bulk to the solid structure beneath it; and in biology there were two new boys, one from Texas, one from Wyoming.

Each day Sheila McRae, a tall girl too lanky for self-confidence, left school and glanced over at the strange building going up in the town in which she had gone to school all her life. Then she climbed on the school bus and rode home to her father's ranch on the Rosebud. But Sheila seldom talked about the power plant at home, and when she spoke in the classroom she preferred to argue about evolution; but the plant was always there, its bulk growing.

Like Barbara Wimer, Sheila McRae went to school with a group of kids she had known since first grade and before. Some years those twelve had been the only students in the class. But now the thing that Barbara's class had speculated upon was happening; the town was beginning to boom. In the fall of 1973 there were almost 40 students in Sheila's junior class. The school was packed. Classes, she remembered later, were even being held in locker rooms.

And the town was altogether changed. Until about 1970 it had been as she had always remembered it: a small rectangular block of streets and houses and trees set on a slight slope to the east, surrounded by hayfields and pasture, with the grain elevator at one end and the school at the other. Now it had filled the little valley. Dusty roads surfaced with scoria reached out across the places where the fields used to be, and in a basin on the east side of a little swamp there was a huge new pad of crushed scoria covered with trailers. There were trailers to the south of town, too, and up to the north the B & R Bar had put in a small trailer park. A few of the people in the trailers had trucked topsoil to their tiny backyards, and had tried to grow gardens, but the occasional little scabs of earth and faded greenery only emphasized the bleak stony field in which they lived. The roots of the people were as impermanent as the little gardens; the village they created out of those acres of wheeled sheet-metal boxes shed an aura of impending flight. Colstrip had become a temporary anchorage for a flotilla of the homeless; the image it created was that these were just a few, a small portion, of a huge national wandering population of the discontented. Like hopeless gypsies they came and camped and went away again in the dawn, and who knows whose children they would take with them on the road?

And at the center of it all, climbing like a stone monolith in a nomad camp, was the one permanent thing they would leave behind, a symbol of their restlessness and their need, the power plant.

"A lot of people in Colstrip were really depressed," Sheila remembered. "All the time."

Then one day there was another new boy in class. He was not as tall as Sheila, he had curly hair and wore baggy blue jeans. On his face was an expression of distaste for what he saw around him, and some of the students mistook it for arrogance. His name was Mike. And one day Sheila found herself walking between classes beside him.

"I hate this place," he said, without preamble.

Sheila did not like to seem surprised. She put on a sardonic air. She didn't think she was going to like this boy.

"Oh, really," she said. "That's nice to know."

"I sure didn't want to come here," he said.

"Where did you come from?"

"Rock Springs, Wyoming. And before that, Centralia, Washington. Before that Oregon. Before that California."

"Oh." The places spread out before her like train stations, each suddenly mysterious, almost alluring.

"Where are you from?" he asked.

"Here."

"Just here?"

"Um hum."

"Never been anyplace else?"

"Nope."

"You were *born* here?"

"Yup."

"Your dad used to mine here?"

She looked away, showed him the pride in her profile.

"He owns a ranch."

"Oh."

"What does your father do?" she asked.

"He's a foreman for Bechtel down at the power plant."

"Oh," she said, then rushed on: "What do you think of evolution?"

29

[*Rosebud County*] *Sheriff Jim Schiffer has advised parents of the county that hard drugs such as heroin, cocaine, and methadone have cropped up in the area. Schiffer said a kit which included all the items necessary for the intravenous injection of the drugs was confiscated last week in the Colstrip area.*

—Forsyth Independent
February 7, 1974

A cold wind blew down the main street of Forsyth, down past the Roxy Theater and the Yellowstone Pharmacy, past the windows of the Joseph Hotel and Café, where an old cowboy sat on a couch with his boots up on a radiator and watched the people walk by. It was early spring in Rosebud County, but there were still little drifts of snow, blackened by coal dust, under the rails where they ran through town.

George O'Connor, president of Montana Power Company, had brought Martin White with him to Forsyth to talk with the county commissioners about the possibility of issuing county tax-free revenue bonds for the construction of a shopping center in Colstrip. Now he and White were just finishing their meal in the café. White had been to Rosebud County more times now than he wished to count, but today he saw it with different eyes. In the years since he had gone full time with Western Energy he had absorbed responsibility into his voluminous capacity to work like an expanding star consuming planets, and in 1972 he had been assigned to manage the town of Colstrip. He had hired a consulting firm to advise him on town planning, and he had begun to direct the growth of the town. But the swiftness of the growth caused more problems than expected, and in early 1974 the company had asked White and his second wife Sheila to move to Colstrip. So as he looked around the county seat of Rosebud County today it was with the eyes of a man who would soon be a resident.

The café was crowded, but not noisy. The Formica-topped booth tables seemed weathered, like many of the men who sat around them. On the walls were black-and-white photos of cows, bulls, and men; the eyes of the men were hidden in the shadows under their hats.

White and O'Connor got up and started for the door. But, seated at another booth, was a man O'Connor knew. He was a lean cowboy in Levi's, boots, wearing a scarf wound tightly around his neck like a collar.

"Say, Wally," O'Connor said. "I'd like you to meet someone."

McRae looked up.

"Hello, George," he said, and looked White over. White looked back at him with his forthright blue eyes.

"How're you doing, Wally?" O'Connor asked. "How's Duke? How's Ruth? And the kids? Have you met Martin White? Martin's in charge of Colstrip for us. Hell of a good man."

They had met once before, for a few seconds, during a hectic legislative session. White didn't quite remember. So now McRae indulged in a gentle cowboy bullying.

McRae held out his hand. White shook it. McRae said:

"*You're* the guy."

"Yup," White said. O'Connor's face was carefully blank. He had known McRae for years. McRae's face, too, showed nothing. His eyes were narrow; the moustache was a tuft of belligerence.

"You're the guy," McRae continued. "You're the guy that's supposed to come down here and visit with me about all the problems we've had. You ever been to Colstrip before?"

"Sure, Wally," White said, "I've been there quite a lot."

"Well," McRae said, "you know, I don't live too far away from there but you've never showed in my place and I thought maybe you would."

"I'm going to be visiting with you," White said. "In fact, I'm coming to Colstrip. I'll be there, and I'll come down. That's a promise." White may have thought that was the end of the conversation. But O'Connor, the gleam still there, didn't move.

"You gonna fly or drive in?" McRae asked.

"As a matter of fact, Wally, I'm moving down. I'm going to *live* there."

"Ohhh," said McRae. "You're going to *live* in Colstrip? Gee, that's too bad. I'm sorry. You know, most people don't like it very well."

White couldn't be sure who was applying more pressure, McRae or O'Connor, who just stood there watching, amusement alive behind his mask of sincere interest.

"I'm sure I'll like it fine, Wally."

"We-e-ll," McRae said. "Don't say it until you've tried it. But I'll tell you: It's gonna be real interesting to have you down there. I'm sure that you may not enjoy your stay, but I'm sure that it'll be *very* interesting." He put out his hand. "See you, George. Good luck, Martin." McRae's eyes never softened, but in his final words, at least, there was a trace of warmth. Martin White followed George O'Connor out of the Joseph Café and into the wind.

30

When Lois Olmstead and her family moved down to Colstrip from Billings on the day after Christmas 1974, in a blizzard, their home wasn't there. It was still being manufactured in a housing factory in Colorado. All that was there under the drifted snow was a foundation. So Lois and her husband, Robert, and their three young boys, packed all their belongings in two garages, moved into one unit of a fourplex, and did a lot of praying. Although Lois Olmstead liked to attack trouble by being aggressively cheerful, it may have occasionally occurred to her that she and her family had been sent to Colstrip by God as a test of their faith.

About the time they were getting settled into their new house, which finally came down the road in March, an underground pamphlet began circulating in Colstrip. It was a little brochure put out by an individual calling himself the R & H Travel Agency. A copy came to Lois Olmstead one day and, feeling it was a reflection of at least part of the culture of this curious town, she added it to the neatly organized collection of souvenirs, photographs, and notes that told the story of her life.

"Have you ever wanted to break free from the rat race of living in *A Big City*?" the brochure asked. "If you have ever felt this way R & H Travel Agency can show you the way to a full and rich life on the edge of the Rocky Mountains in Colstrip, Montana." It went on to list some of the attributes:

Living accommodations available range from clapboard prefab homes, hotbox trailers, to open range.

Colstrip ... enjoys a truly healthy climate. From sub-zero temperatures in the winter to 100-plus degrees in the summer (the average is pretty good!). Scenic wonders and barren landscape abound under partly cloudy skies.

Medical Care: Professional medical care is available two hours each week. Hospital emergency room care a mere 100 miles away.
Entertainment: Two local bars feature a wide selection of beer (from all over the country). Firearms are permitted but spurs must be removed before entering.

On the back of the brochure was a call to work:

Local industry is looking for a few good men! Interested persons may contact the R & H Employment Agency for details. There is always a job opening in Colstrip, Montana.
We need you! (To get us out.)

In the spring of 1974 the mood of Colstrip turned bitter when the members of Operating Engineers Local 400 went on strike, arguing for higher wages and higher subsistence pay. Other trades honored the picket lines.

"I guess it was easy come, easy go," Lois Olmstead said later. "There were people who had been making thirteen dollars an hour who suddenly didn't have food to eat. There were people who were destitute, penniless, and we'd get together and give food. Then it didn't matter who you were. I remember everybody was giving food and money. I'm sure it did something to the ranching community, because they could see what it was like for people to become enemies over a strike—I'm sure it confirmed their fears: that we don't want business, we don't want this kind of thing here."

The strike came to a conclusion sourly for all sides when the wives of other workers who were honoring the lines picketed the picketers, demanding a settlement so their men could go back to work. A week after the counterpickets showed up the strike was settled.

But if the strike was the kind of drama that Lois Olmstead, whose life had included tours of duty with her husband on several other construction and generation projects, understood, the encounter with the Indians wasn't. The Northern Cheyennes, although a few were employed in Colstrip and Myron Brien and one or two others lived there, were not a part of the social structure at all. They were more of a mystery, living inscrutably in their reservation towns, or driving through Colstrip in old cars. To the instincts of the townspeople their cultural distance made them more of a threat than were the ranchers; you could understand the ranchers' defense of property, but the Indians were strange and unpredictable.

So the night that they gathered at the B & R Bar just an arrow's flight from Colstrip stunned the town. It was the night before Easter. Lois Olmstead recorded it in her diary.

"Just finished doing the Easter Bunny bit hiding eggs for the three boys plus my niece and nephew and Mom and Dad, and Mom and I hit the hay. Robert groggily reached for the bedside telephone when it rang at 2:30 A.M., and we both sat straight up in bed when we heard the first sentence. 'Bob, there's Indians gathering. We expect trouble. The sheriff just called and the American Indian Movement has been arriving in here all day and there's about two hundred of them drinking up at the B & R Bar.'

"Right here I'm thinking, 'We've got to circle the wagons!' That's all I remember from cowboy and Indian days.

"Robert has his pants on and is buttoning his shirt while we hear one of the guys say 'We think you better get your power company car away from your house in case they find out you're there, and take it to the plant and away from your area.'

"As he hung up, grabbing his shoes and car keys, I could see the bedroom lights of the four other owners of company vehicles come on and activity in the homes on our street. When Robert left with me giving shaking 'Be carefuls' I went to the guest room, explained the unreal phone call to my folks, and they got up, too. All of the recent kidnappings came to my mind, and with Robert being plant superintendent we locked all the doors and went to keep watch at the windows. Looking out we could clearly see the aroused power company families because their lights were on and right away we decided this could be a bigger giveaway than the cars so we shut all the lights off and walked around in the dark.

"The other wives immediately shut their lights off. We kept thinking of Wounded Knee and the three or four weeks of roadblocks. I was thinking of National Guard, of how glad I was of a full pantry, and where was my husband at that time.

"Dad said tonight of all nights he didn't want to get stuck in Colstrip.

"We had nervous jokes and whispers until Robert returned in an unmarked car an hour later. He said all various construction heads had locked up all the dragline gates, the mine area entrances and doubled guards at the plant construction site, so there was nothing we could do but go to bed and they'd call him if there was trouble. And me thinking what if they cut the phone lines.

"He sounded calm and confident. I followed him into the bedroom where my courage quickly vanished as he removed his gun from the closet, loaded it, and put it near the bed. He was soon asleep and I lay keeping guard, wearing my glasses lest I mistake the door for an Indian moving in.

"Morning kept its routine of arriving in daylight, and the night before seemed a long time ago. Yet as we came home from Easter sunrise services the padlocks on the gates that up to this time had been left open let us know that there was reality in the night.

"The kids found the eggs and baskets and as we talked of the night before they were unaware that during the night their mom and dad had played real cowboys and Indians in Colstrip."

In between these unusual diversions there was too much time in Colstrip, particularly for the women. Or it was not so much the time as the emptiness it framed. You could eat at the B & R Bar, or the Arrow Café—a trailer parked down near the power plant—or have a pizza at the Frontier Village, where one owner of the parlor provided some momentary diversion in March by killing himself in a homemade airplane. Or you could drink. To shop, go to a show, or even get your hair styled you drove to Forsyth or to Billings or to Miles City. There was too much time, time that, aided by the winter snow, spring mud, and summer heat, sealed them in their trailers and prebuilt homes while the men's shifts swung in and out of darkness like the lunar day.

"Everybody was working," Lois remembered. "And you never knew when the husband was coming home, because it varied with the shift. It was a town of never knowing when anyone was off or on."

Lois Olmstead's native enthusiasm made her almost immune to the boredom of Colstrip. She found diversion in unlikely places, but she found it. One of those places was books. At one time the favorite book among the people she would call "wives of supervision" was a novel called *Camerons,* by Robert Crichton. For a while its catch phrases were passwords into a little literary clique, but whether the parallels that the women saw in the book extended beyond calling the Colstrip store the "Pluck Me" and the transient trailer court "Doonietoon," Lois didn't say. It's possible that the grim conditions of underground mining and extreme class oppression described in the book gave the women a chance to laugh at their own less violent problems while they drew lighthearted parallels, but it is also possible that as Colstrip blossomed into a mire of construction surrounded by the pale raw hills of spoil, the women of the town found some passages in the book too familiar.

The town looked small from where they were, and black. That was it, black, a stain on the moors and snow surrounding it, but a place to make money in. . . .
Coal corrupts, Gillon thought; the very nature of the act, ripping open the earth, disemboweling it, stripping its black veins. Coal corrupts and mining coal corrupts absolutely. . . .
He did get used to [mining] which later he would see as a crime against humanity and reason. Man accepts too much too easily, and he learns to accept the unacceptable.

Although Martin White later denied that there was any attempt to segregate Colstrip residents by class, the effect of certain Western Energy and Montana Power Company policies of locating people according to their times of arrival in Colstrip and sometimes by their employer created relatively distinct groupings. In general, construction workers lived among the stones of Burtco Court; more permanent employees of the prime contractor, Bechtel Corporation, lived in the southern, more permanent trailer park, and Western Energy and mine employees lived in the old part of town, in boxy houses on rectangularly arranged streets, and people who worked in the power plant lived in the new houses to the north. These groupings corresponded roughly to "Doonietoon," "Uppietoon," "Moncrieff Lane," and "Tosh Mungo Terrace" in *Camerons.* In the book the different groups sat apart from one another on Sunday in the Pitmungo Free Kirk; in Colstrip they could be seen in less formal arrangement at Colstrip Colts basketball games. In an odd additional twist, Lois remembered, "The supervision people tried to be in the ranchers' class and get in with the ranchers who were making a name for themselves, and call them part of the structure of the social class."

And yet in between them stood the power plant, huge, silent, ragged still with scaffolding and surrounded by construction debris, and so far standing still and motionless in the valley, a skeleton machine without its voice. But to Lois, whose husband was plant superintendent, it seemed no obstacle at all. The real barrier between people like herself and the ranchers, she felt, was simply a lack of communication.

"On the tours of the mines, you know, or at meetings, they would often have my husband speak, then maybe Martin White, and then Wally McRae," she remembered. "And they'd say let's get the three different opinions—never let's see what they have in common." In the reports, hearings, and newspaper articles, it was always conflict, not resolution. When Montana Power would be accused of trying to get around pollution-control rules and Lois "couldn't see my husband wanting to put all kinds of gunk out of the stack so it would blow down on his family," she often thought, "If only we could just talk together, we could solve it."

The Olmsteads themselves were communicating without restraint across the lines. Shortly after they moved to Colstrip they had met Nick Golder, one of the more outspoken ranchers, and had become friends through their common preference for fundamental theology. "We came from a background," Lois Olmstead said, "where you packed your Bible and you read it." The community church "did not meet our specific needs," so they started meeting for Bible study weekly. The two families soon formed the nucleus of a group that included representatives of most Colstrip classes. It was this free mingling under the common umbrella of Christianity that gave Lois Olmstead hope.

She felt it most strongly on those afternoons when they would go down the road halfway to Lame Deer and pull into Nick Golder's yard, where the big old house nestles in trees and looks out across the creek into the red-topped hills. On those afternoons they would gather in the big room around the organ and talk about God. They would pray for sick cows or for the strikers down at the plant. Once Nick got his barber to come down from Billings to share his personal experiences with Christ. There was the rancher, the superintendent, a welder, a laborer, an electrician, a union man, but "nobody was anybody there. We were together because we all wanted a closer walk with the Lord." They would cluster together and sing hymns: "I Have Decided to Follow Jesus," "How Great Thou Art," "God Is So Good." The voices would glow faintly out into the pink and amber spring evenings on the Rosebud, sounding, out in the deep, soft corrals and on the baked rock of field and hill, as thinly and sweetly as the distant longing for reconciliation among human beings sounds to the world. It was the music that only a person like Lois Olmstead, clinging with joyful obstinacy to hope in the goodness of man and God, could capture and hold, and take back home with her to the town that stood in the shadow of the power plant.

31

In and around Colstrip . . . several families are moving out of the area and going to Alaska to work on the pipeline. . . . Mike Brosios left Sunday for the North Slope. . . . His wife, Sharon, and their children will be packing their belongings to send by barge and they will fly to join Mike as soon as their mobile home is sold. . . .

The Rosebud Protective Association appealed to the companies and to the legislature to "prevent the local subsidization of the impact [on schools] caused by rapid industrialization of rural areas. . . ." The landowners group feels that Montana Power's offer to provide some mobile classrooms this fall is no real solution. . . .

—Forsyth Independent
May 16 and May 23, 1974

Colstrip was a chrysalis of mud. The streets were mud, the yards were mud, and where construction was going on there were heaps of mud piled in neat rows, as if waiting to replenish places in the town that might accidentally dry out.

When Martin White went to work on the first day he was in Colstrip, April 4, 1974, he spent most of the day answering phone calls from irate residents who wondered when he was going to do something about the horrible living conditions in his town. By the time the day ended he had an idea why the company had decided to send him to Colstrip. There was an immediacy to the problems of the people here that did not get through in its full intensity when you lived in Butte. When White went home that night, ready for some peace, he found his wife, Sheila, standing with her mother out in the yard, knee-deep in mud, unsuccessfully trying to build a dam to keep the water out of the basement.

119

It was not much better in May. Repair to the muddy town was sluggish; the houses that were there were now relatively dry, but many of the factory-built homes that should have been there had not yet arrived, and where yards and lawns were supposed to be growing there was nothing but earthen fields scarred by rivulets and the small craters inevitably left by playing children. Much later White would write wryly in a report that "complaints seemed to be in inverse ratio to the quantity of green grass in residents' yards; the more green grass the fewer complaints." In May, there was no green grass at all.

The weather was scuddy; low streaks of cloud snarled over the raw town and fed the general discontent with bursts of wind-driven rain. Leaves were beginning on the trees in the old part of town but they were hesitant and pale in a land that did not seem to be done with winter.

On an evening in May, Nolan Fandrich, one of the few men in town who had worked in the Northern Pacific operation, was talking with Martin White in their shared office, which was the bedroom of one of the small residences in the old part of town. White was examining the last papers of the day before driving the two miles over to Burton Farley's ranch house for a meeting with the area ranchers. There was a lot of bounce in him tonight: confidence or adrenaline, or maybe a little fear.

When the topic got around to the upcoming meeting Fandrich gently shook his head.

"Looks like it'll be a bad one," he said.

"Yep," White said briskly, with a grin. "If I'm not back here by ten you'd better come looking for me."

The two men laughed.

When Martin White lived in Butte and worked at 40 East Broadway, the ranchers of the Colstrip area had been a distant irritation, a minor political nuisance, pesky but off the centerline of business—the way Indians were a problem to Lincoln during the Civil War. Now he was in for an education. "When it came to some of the things that went on around here," he remembered, "I was as dumb as a wooden watch."

At the time he went down to Burton Farley's to talk to a group of ranchers about the plant, schools, and Colstrip, the Rosebud Protective Association had just been formed and had resolved, as a first order of business, to demand that coal be shipped away to generation near West Coast power users if it was to be mined at all. More ranchers like

Duke McRae—men who were slow to anger but hard to bluff—were mobilized now, by the increasing population of the school, which had risen 52 percent in six months; by the evidence that hard drugs were in use in Colstrip; and by the threat of the power plant. There was no distance to filter their anger when he went to the meeting at Farley's home.

The room was already full when White arrived. It was a basement recreation room, low and long, with a hint of an adjacent furnace, and tack hung along the walls. The ranchers and their wives were arranged without symmetry on a collection of unmatched chairs, sofas, and benches; one leaned against a saddle that was thrown over a sawhorse in a corner. White guessed there were about thirty-five people there. Sniders, Farleys, McRaes, Gillins, Streeters, Egans. Their pale foreheads gleamed; they were a hardness, like a field of quartz, in the room.

White stood, his blondness and his muscular athletic build in contrast to the dark Scotch faces and lean cowboy lines of many of the men. His back was to a small upright piano. In front of him was a skittle pool table, a barrier or a challenge, between him and the men and women who waited, hostility checked, for what he had to say.

"I suspect it didn't take me more than ten minutes," White remembered. "I just basically gave them an idea of what I knew about what we were going to do—what we were going to do on the town and what we were going to do on the power plant."

But perhaps White's simple story didn't ring as such to the group.

"Martin came to speak instead of listen," Wally McRae remembered. "He came to our community with what we considered was a dearth of information and started to tell us about our community. And that our perception of our community was wrong."

When White finished there was a moment of silence in the room, a mustering of energies and force among the ranchers, a rustle, a squeak of chairs. Then one spoke. He was off to one side, an older man.

"If you were George O'Connor," he said, "I'd get my shotgun and kill you."

White looked at him. So it was that way. He did the right thing; he didn't back down.

"I think that's a damn unreasonable thing to say," he told the man. "And I personally do not feel that I came here to listen to that kind of stuff, and if that's all that's going to occur at this meeting I'm going to leave right now."

There were no similar remarks, and White's aggressive recovery was respected by the group, but the evening remained tense. There was talk about the condition of the roads, and about whether Three and Four would just be the beginning of a sequence of plants that might stop at Unit Ten, and there was talk about schools, and that's where White got into trouble.

"The Colstrip school district has had to bond itself for an extra $750,000," said Duke McRae.

"Things are going down while you're telling us they're getting better," said Nick Golder. "And this doesn't give much credibility to what else you're saying. While our taxes have gone up the quality of schooling that our kids are getting has gone down."

And that irritated White. Montana Power Company was planning to give eight modular temporary classrooms to the school to help take care of the expansion until the property taxes that would be levied when the plant was completed could take up the slack; there was some sentiment in the company against the construction of larger facilities which might then only be empty when the plant's construction period ended. And construction was almost by definition a time of turbulence. All that was needed to survive it was patience. The school district would get its huge cut of taxes soon enough, and then the system would be better off than ever before.

"Look," he said. "Our investment in the plant will be $183 million. Our investment in the town will be $10 million. Our investment in mining equipment will be $24 million. There will be sufficient tax base here to build a gold-plated school."

There was the error. Not the statistics, but the last phrase. It held the ring of arrogance when probably the only thing that drove it was frustration.

"It was a *horrible* thing to say," McRae remembered. "And that really turned people off. There was a chorus of answers: But Martin, we don't want a gold-plated school."

So White learned the anger and the sensitivity of his new neighbors. And next time he went to Butte he told Paul Schmechel: "The biggest impression that I left the place with was that they were sincere about this and this wasn't some little game they were playing."

But when White drove back to Colstrip that night, after ten, it wasn't exactly like returning to a sanctuary. The people of Colstrip, who were mostly construction workers, were even less patient about the temporary afflictions of the construction stage than were the ranchers. So he was under siege at home, too, and in some ways the tactics his ideological allies used there were less straightforward than those his enemies used out on the range.

White had remarried before moving to Colstrip; his wife Sheila had to face the problems of development debris like all the rest, but with an additional twist: She and her husband were the ones blamed for it.

The discontented accosted White at work, on the street, and at home. Why did the rent go up? You've got to do something about this mud. You said I'd get reimbursed for finishing my basement—where's the money? Once White almost had to throw a couple out of his home. "They were going to fight us and work us over," he remembered. "They left; I didn't have to bother to throw them out. But Sheila was there—Sheila is a little bit of a person—Sheila was in there filling a pan with hot water; she said she was going to pour it on them."

Sheila White is a small, restrained woman with precisely pretty features, cool as porcelain. When she came to Colstrip as the boss's wife she found herself drawn from anonymity to near infamy in one move. "Six months before I moved here I was a smelterman's daughter," she remembered, "and now they were trying to treat me as if I was an O'Connor."

For several weeks Sheila tried to ease herself into the role of the community leader's wife. At least once a week she invited several wives of new and older Colstrip residents to teas in her home, hoping the occasion would promote new friendships and ease the pervading discontent she could feel in the town. She hosted with a cool friendliness she hoped might fit the role forced on her. But suddenly the teas turned to disaster.

One day only two women came. She had invited at least a dozen, but the room had remained virtually empty, the two women who came sitting absurdly among an assortment of cakes and small sandwiches made for twelve. And when Sheila looked outside, a terrible fear shook her. She suddenly realized it would take more strength than she knew she had to cope with Colstrip. The role that she had so grimly accepted, that she had tried to assume with grace, had been thrown in her face. The women she had invited were all standing out on their lawns, talking to each other and looking down the street at her home.

"And Sheila just withdrew," White remembered. "What had

happened, there had been a fellow, he and I had had a little disagreement or something at work, so his wife called up all the ladies and told them not to go to the party. So Sheila quit having parties, and she virtually closed the door. It was a bear for her. It was a hard time. Because for a while the worst person in town was Martin White. And we weren't accustomed to that."

In the midst of turmoil, the power plant thrived. Its roots were down, its superstructure complete. Lights shone in its cavernous interior day and night. At last the crews began to assemble the huge stacks, the five-hundred-foot-tall vents through which the plant would exhale and blast its polished, faintly sulfurous breath across the plains.

32

It was the opinion of the Rosebud Protective Association [at a Chamber of Commerce meeting] that the decision of whether to construct Colstrip Three and Four will mark the watershed between all-out exploitation of the area as a power generating center, and limiting the power generation facilities. . . .

"It's not going to end with Three and Four," [said Wally McRae]. "There's going to be Five and Six and a gasification plant. They're going to keep it up until there is resistance."

—Forsyth Independent
June 20, 1974

The irons stood glowing in the blue-and-white flame of the propane torch. Duke McRae squatted beside them, his shirtsleeves loose at the cuffs, his ragged and sweat-stained straw hat shading his face, waiting. At the edge of the corral stood calf wrasslers, two teams of young men, waiting. At the fence a man holding a syringe squinted at the needle into the sun, and pressed the plunger to produce a small bright globule of three-way blackleg vaccine. Beside him Wally McRae slowly stroked a knife against the gray, weathered surface of the top crossbar. The air was clear, smokeless. There was little dust.

The scene was almost a still life, a picture of men preparing to brand. The only action was in the center of the corral, where the roper spun his horse, built a loop, threw it, and missed his calf again. One of the young calf wrasslers worked his bulging lower lip for a few seconds as if he had caught a mouse in it, then spat brown juice on the ground. Wally McRae whetted the knife on his pants' leg. Duke McRae moved the bottle of propane to direct the flame more efficiently on the irons.

The roper turned his horse again, and the hooves thumped in the soft earth.

Duke McRae looked up at the roper, a little anxiety in his face. It was the new kid from Kansas, Charlie Wallace. McRae knew how much he wanted to do well.

Charlie Wallace had married Barbara Wimer in January 1972, and they had moved to the Wimer ranch. Wallace had always liked to be close to his family, and when Barbara married him some of his friends warned her never to try to take him away from Topeka. But when he got to Montana, Barbara remembered, "It was like he'd been here all his life." Wally McRae remembered it similarly. "A lot of people can live here all their lives and never become a part of it," he said. "But Charlie was born to be here. And it just took him twenty years to find that out."

Charlie Wallace was lean and energetic. His philosophy, Barbara remembered, was simple: Work hard and play hard. In his favorite photograph, the one Barbara chose to have printed up and sent to his friends two years later as a memorial, Charlie is astride a horse, holding his rope and looking quizzically past the camera, as if speculating on the character of a bull that awaits him just out of the picture. In the photograph his face shows the pale forehead and tan cheeks that are the mark of a man who spends his days working in the sun under a hat.

"He was very open, up front," McRae remembered. "If he didn't know anything, he didn't; he wasn't embarrassed about asking. And he learned, quickly."

"You could tell him how to do something or he would watch someone do something and then he could do it," Tom Wimer, Barbara's father, said. "And he always done it the best possible way."

Charlie and Barbara lived up at the Wimer ranch north of Colstrip. Every day he would drive down past the town to go to his job as a ranch hand on Duke McRae's outfit on the Rosebud. Every day he would pass the power plant and he would glance across the mass of unfinished houses and trailers that was Colstrip, at the huge skeleton of steel and concrete that stood against the horizon. If it bothered him at all he seldom spoke of it; sometimes he told Barbara he knew how few open places were left in the United States and he wished they wouldn't clutter this one up with trailers and machines. And he was as disturbed as the rest of the family when the mine claimed a beautiful sandstone formation on Cow Creek where they used to go on picnics. But he didn't object when Barbara went to work as a secretary for Bechtel Corporation, the prime contractor on the power plant; in fact, he took advantage of that contact to play on the Colstrip Bechtel community basketball team. He was too enthralled with the mystery and the power of the life of the cowboy to believe it could be seriously threatened.

And now he had achieved a special recognition: After two years of learning the intricate differences between rodeo and the real thing, he had been asked to rope at a branding. Here was acceptance, the final gesture of inclusion in the rites and ritual of the cowboys of Rosebud County. Charlie couldn't hope, because of his city background, to be a part of this country's rich past, but this made him a part of its future.

And yet, here it was—the branding—and he couldn't rope a calf. In how many rodeos had he thrown the rope, jumped from his horse, whipped three legs of the calf together with a piggin' string and flashed his hands high, and all in less than fifteen seconds? But here—where they were all waiting, watching, saying nothing, in the clear hot air of a spring morning on the Rosebud, here, where it mattered most of all and the prize was not money, points, or a belt buckle but entrance into a way of life—here the rope twisted in his hands, became unruly, and snaked away from the calf—again, and then again.

He wheeled the horse. The horse fidgeted and backed. He built his loop again and felt the faces, some watching, some looking at the ground or at the layer of thin, high clouds that fanned out from the west. Then Duke McRae looked up at him from his place by the irons.

"Why don't you fight your horse," he said mildly. "That'll help. Seen a lot of fellows do it."

Wallace laughed. There was a little chuckle of appreciation from the older members of the group. Wally McRae, alive to all the nuances of the life he cherished as much as Charlie did, thought, "Good for you, Duke; so we renew the tradition. It survives another generation." And Wallace relaxed. He turned and built his loop and shaped up a calf and got him and brought him in. The wrasslers jumped to their work. The calf fell. Duke McRae grabbed an iron. Men came with the knife and the syringe. The acrid smoke leaped up into the still air and curled gently away over the heads of the branders. And by then Charlie Wallace had another calf. Nothing could stop him now.

33

About the time Clifford Powell noticed he was being shot at, he seemed to be already running. He could remember before he took off seeing a man lying on a rock up the hill, and now he could hear the sound of each shot and the faint hum of the bullets going by somewhere overhead, and he had a hopeful suspicion—not enough for comfort—that perhaps the shots were aimed above him and were not meant to kill.

But he kept running, pumping up the dirt road, until suddenly in front of him he found a motorcycle parked across the road, and another man, young, tall, lean, his face shadowed by a wide hat, leaning against the machine.

"What the hell do you think you are doin' here?" the man asked.

Powell might have asked himself the same question. He was a pipefitter, a forty-five-year-old member in long standing of the International Union of Plumbers and Pipefitters. His burly form was rounded at the edges now, and as he stood in front of the man on the motorcycle like some kind of juvenile hauled before a hanging magistrate, he puffed slightly from the short run. He had large eyes with bags under them, and from them now shone very little hope.

Powell was employed as a general foreman in the construction of the power plant at Colstrip, and he lived at a trailer court about three miles south of town. If someone had come to him at the B & R Bar and bought him a beer and asked him what he was doing down here in eastern Montana, in general he might have said, filling out a little with pride in the men he worked with as much as in himself: "You've got to have energy in this country and Montana has an obligation to furnish its fair share of anything we can do to strengthen our country, and wherever you've got to have energy you've got to have people like us that are willing to do this kind of work." But the question of what he was doing on this particular piece of ground confronted by what appeared to be a crazy cowboy bent on violence, was less clear.

Powell commuted home to Billings on weekends, but he occasionally liked to drive out into the country on weekdays after work. "There was always some deer or antelope out there to see," he remembered. Today he and another pipefitter had driven up a dirt road that leads east off the pavement south of Colstrip. Partway to the Rosebud he had met an Indian walking toward town from his stalled car, and he had picked the Indian up and gone out on another, smaller road to see if he could help fix the car. It was then that it happened. The three men had just started to work on the car when he saw a gleam of metal out of the corner of his eye.

"I noticed over in some big rocks that were there," he remembered later, "I noticed a handlebar of a motorcycle sticking up—the sun was shining on it—and doggone I looked up and there's a guy lying on a rock with a gun, so I took off as fast as I could. Well, the guy shot at me three or four times—I don't remember exactly—so I zinged up on the road and there was this other one sitting on a motorcycle right in front of me."

"What are you doing here?" the man repeated. His eyes were almost invisible in the hard shade under his hat.

"We were coming to help this—"

"I don't really give a damn," the man said. From the corner of his eye Powell could see the other man walking slowly down from his gun platform on the rock, carrying the rifle with the lazy confidence of a man who has just shot a deer in the jugular at two hundred yards. "We know who you are," the first man continued. "We know you're from the steam plant."

There seemed to be nothing much to say in answer to this; as an accusation it seemed, to Powell, to lack a criminal element. But he was uneasy. He had dealt with hard men before, both on and off the job, but these were unfamiliar types. They were like caricatures of television villains, almost that absurd, terrible amoral cowboys with sweat stains on their hats, tobacco packed behind their lower lips, and sadism gleaming in their eyes. If he was lucky they would do something like take his boots and his car and make him walk home barefoot.

"We want to tell you bastards from that steam plant something," the man by the motorcycle said. "We want to tell you that this is private property—*private* property—and we don't like to see your ugly faces on our land. You bother the cows. Now get your asses out of here."

The Indian had disappeared. Powell and his friend walked back to their car, turned it around, and drove slowly south along the Rosebud. The scoria road led on and on into the dusk. The two men said little. Once in a while they passed a ranch. The windows seemed to stare at them with enmity; the men in the corrals looked up as they passed. They did not wave. Powell and his friend met someone from the County Sheriff's Department on the road, stopped him, and made a complaint. "Um, hum," said the deputy. He went north, trailing dust. He seemed to be smiling at something when he left.

At last the two men turned onto the paved road up to Colstrip. The road here was swift and sure. They came up over the hill and there was the town, lights gleaming. The power plant rose, huge and brilliant on the horizon, like a friendly fort.

34

The old prophecy of the Cheyenne Sweet Medicine was self-fulfilling; it carried the makings of its own truth in its despair. "They are coming all the time, to turn the land over and kill it. . . ." The St. Labre Indian Mission in Ashland called the Cheyennes the Race of Sorrows, and all the prophecy spoke of was a slow and weary ending. For almost a hundred years the tribe was caught in this trap, in an awareness of the inevitability of defeat, fed from within by the prophecy and from without by the casual lack of consistency with which the United States government handled the tribe it was supposed to protect. And there was a sweet kind of comfort in the promise of racial doom: It banished all the terrors and uncertainties implicit in a clear future and let the tribe steep in the present's slow decline. The sale of coal leases opening half the reservation to strip mining had been just a final expression of that comfortable despair.

By the time he went to work in the Western Energy mines in Colstrip, Myron Brien, a forty-two-year-old Northern Cheyenne, was sick of all that. For most of his life he had lived on and off reservations, working in seasonal construction or as a tribal police officer. He had worked on the Northern Cheyenne reservation and others in the Dakotas, and most of what he had had to deal with was racial despair surfacing in drunkenness, fighting, and pervasive pessimism. To him there was no magic in being of an ancient North American race.

Brien was tall. His body had not gone to fat, but he carried a small pot belly around in front of him like a deflated basketball. He liked to wear jeans and long-sleeved cotton and wool shirts, and he was seldom seen without the tag of a Bull Durham tobacco sack hanging out of his shirt pocket. His face was long, dark, and deeply creased; it looked as if it had been carved of old rock stressed by time and

131

pressure. It was wholly without conviction; although he had at last rejected the reservation and the tribe, his face was as its spirit, weary, and short of hope.

Brien was a heavy equipment operator. He drove front-end loaders, bulldozers, scrapers, coal haulers, and water wagons. They were the intermediate machines in the mine; standing alone each would seem gargantuan, wearing tires a man could hide in, wielding blades that could carve a three-bedroom home into matchwood in two passes, or carrying beds that could hold the entire harvest of wheat from seventy-five acres in one load and dump it in twenty seconds. But next to the real heavyweights of the mine—the draglines that peeled back the earth and the shovels that scooped out the coal—they were as light and agile as kittens playing around an old St. Bernard. Brien would have liked to operate one of these great vessels of the mine, but he was satisfied with what he was doing. The money was good.

"Hell," he remembered. "My net take-home pay when I worked over at Ashland as the police officer was something like $82.60 a week. But I'm making as much now, gross, in one day, as I made a week over there."

So he was a part of the industry that was poised to enter the reservation and mine it for its coal. And if the chance had come, he probably would have helped strip the reservation without complaint, pushing the soil back, hauling out the coal. It was a part of the prophecy, part of the acceptance of the end of the Cheyenne race. He would turn the land over, he would carry its core to the power plant and watch it burn. There was no sense fighting it; it was destined to come.

"There was hassles in this country since the time beginning," he said later. "It started out the cowboys and the Indians, then cowboys fighting homesteaders, then along came the old farmer boys, they fought him hoof and nail. There's been wars over that, and he's here. All the fighting, bloodshed, and everything else in the world didn't stop it. And I say the same thing'll come to pass with this situation here. It'll happen."

35

Until 1974 all the Northern Cheyennes possessed of value was their land and their history, and they had sold the integrity of the land. But the history was still there, a story of passionate determination that had wrestled one small blessing from the United States government—the right to live at home. To Myron Brien this was not enough; the subsequent years had extinguished all past glory with defeat. But to Bill Parker the story of the walk back in 1878 shone across the despair of nearly a hundred years and lit a present that held new promise. He had a feeling for symmetry in history, and he believed that another important time for the Northern Cheyennes was coming.

Bill Parker was one-eighth Northern Cheyenne. He could have forgotten it, hidden it, packed that strain of blood away as a souvenir or a novelty, like a double-jointed thumb or a grandfather who was a pirate. And he had had the opportunity. When he was born in 1937 his parents lived in Lame Deer, but his family moved to Billings when he was six. Later he had had the chance to bury his ethnic oddity in the turbulent minglings of cultures and races of Los Angeles, where he worked both as a printer and, much later, in a plant that specialized in cleaning rust from steel. But he returned to Montana when he was in his mid-thirties and began to fulfill his longing to be deeply Cheyenne.

"I'm well accepted by the full bloods," he said once, "to the extent where if a full blood is talking to me he might say something like, you know, 'those *breeds!*' And he looks at me not as a white man or as a breed but as an Indian, see. And this to me is probably the major accomplishment of my life, see, to have that acceptance."

Bill Parker had been back on the reservation in 1973 when the Northern Cheyennes had made the first decision of a sequence that would bring them back into the light of contemporary history. He was a part of the change of mood on the reservation that slowly swung from the racial despair that Myron Brien fled yet carried with him, into something less quiet and more powerful.

The coal leases the Northern Cheyennes had signed at the advice of the Bureau of Indian Affairs between 1966 and 1971 gave six coal companies and speculators the rights to the coal under about 56 percent of the reservation. This leasing would have significantly improved the financial condition of the tribe.

"It would have given us tribal operating revenues in excess of anything we had ever experienced," a Cheyenne report said later, "including a million-dollar health facility and money left over for large per capita payments to every man, woman, and child on the reservation." But, the report continued, "by 1973 the tribe had come to realize that our coal resources were both larger and worth more per ton than we had been led to believe, and had come to realize also the adverse social and environmental implications of the leases." So, seeking to escape once more from the prison of automatic submission into which the government had put them, the Cheyennes armed themselves with a law firm instead of Winchesters and broke out. In 1973 they petitioned Interior Secretary Rogers C. B. Morton to cancel the leases on their land, citing various illegalities in the way the Bureau of Indian Affairs handled them.

During the waiting period between the request and the answer, Bill Parker was one of those studying the whole issue of coal for the tribe. It was this that led him to his eventual conviction in the resonance of history of the tribe.

"I was doing some research on the price of coal," he said, "the coal operations around the country, the export situation in coal, the horizontal, the vertical interlocks, the whole business, see—who would be working against us, who would be coming from what part of the country and from what direction of the general terrain in effect to attack us—and the more I began to look at it the more spooky it got. It became obvious to me that this thing had happened before."

In 1974 he began to write down some of the parallels that struck him.

"In 1873," he wrote, "the people who were my ancestors lived in the vicinity of the Black Hills to the west and south of here. But primarily [they] had their thoughts and religious life attuned to religious practices and worship which were begun in the Black Hills. At the same time officials in Washington, D.C. were wrestling with the financial panic and the stock market and the country was attempting to assemble the last threads resulting in the last push of expansion that finally spanned the continent. And the Cheyenne people were pushed headlong into activity which would be necessary for them to protect their homeland and their way of life: An invasion of the Black Hills, which had been given to the Cheyenne people together with other tribes at the treaty of Fort Laramie in 1868, began. People packing through the Black Hills had found gold and this gold was needed to stabilize the economy of a country badly shaken by war, graft, corruption, and financial panic.

"It sounds like something we recently heard of.

"Today, one hundred years later, one cannot help think back to that time and wonder what went on in the minds and hearts of those who were responsible in Washington, D.C. for this horrible travesty of justice and the conditions that lay throughout the whole land as the aftermath for the Indian people. One wonders if that state of mind does not exist today, so evident in the lessons of history in the halls of Washington, D.C. and in the minds and hearts of people engaged in industry and commerce throughout the land as they proceed to take further advantage of what God has given us, in ways that are far more disturbing and terribly destructive that are to be used in yet another assault on the Northern Cheyenne nation. We know beyond a shadow of a doubt the Northern Cheyenne people, as with the gold in the Black Hills, as with the buffalo, so now with the coal in the Powder River Basin, are asked to offer ourselves up as the first sacrificial victims to this unwarranted invasion upon our land."

But in 1878 the Cheyennes had not gently relinquished their land at the request of the government, and now they had decided to fight again. And again they appeared to win a small victory. In the late spring of 1974, Interior Secretary Morton agreed to suspend the coal leases pending renegotiation on the basis that the Bureau of Indian Affairs had not been acting in the tribe's best interests when it had made the negotiations.

"The tribe and the coal companies," Morton said, "may be assured that the terms and conditions upon which mineral development may proceed on the Northern Cheyenne Reservation will require their joint agreement and support prior to any further approval by me."

"We don't renegotiate with the companies until they tear those leases up in front of us and burn them," Allen Rowland, tribal chairman, was quoted as saying in the Cheyenne report.

"We would be selling not only coal, but we would be selling a way of life," said a councilman, Ted Risingsun.

"This," the report continued on its own, "the Northern Cheyenne are not willing to do no matter how much money is offered nor how badly it is needed."

There was celebration on the reservation. Here at last was an interruption in the deterioration of the power of the Northern Cheyennes over their own futures. At financial cost and perhaps some future hardship they had wrestled the fate of their lands away from the companies that would mine them. Now they could begin again. But to Bill Parker the rhythm of history was not complete; the symmetry his instinct craved was askew. There indeed had been a victory, perhaps smaller and less costly than the one wrung from the tight fist of the United States in 1878, but a victory nevertheless. But it was not yet complete. The land was won back for the time being, but the first phase of the power plant was nearly finished at Colstrip, fifteen miles upwind, and now there stood beside it the threat of the second phase, a machine of twice the proportions. And it had not yet been a full century since the Northern Cheyennes' long, decisive walk had begun.

36

The environmental impact study on Colstrip Units Three and Four is 2,000 pages long. Public hearings on the project have been scheduled, including Forsyth on December 18. . . .

—*Forsyth Independent*
November 21, 1974

"It wasn't hard to get organized as far as this was concerned," Duke McRae said once. "Because it was something that nearly everybody was involved in. But sometimes it's a little hard to *stay* organized. If there isn't some controversy going on, why, sometimes you just get too busy."

But Don Bailey seldom got too busy to forget the anger that had driven him for years. If sometimes he concentrated too deeply on his purebred Hereford bulls or on clearing a patch of land for hay ground, it would all come back to him when he drove up the hill past Colstrip on his way to Forsyth. There, as he rode over the last hill, was the power plant. Now, as the summer of 1974 passed, the plant was nearing completion: The bulk of it would get no higher; one of the twin stacks was finished, pointing a deceptively slender finger at the sky; and the second was halfway up, its truncated top encased in a latticework of scaffolding.

"You leave here and you're driving along thinking about something else," Bailey said, "and as soon as you get to the top of the hill over there, *there it is*." But the power plant that was there, waiting to be finished, was no longer by itself. Now it was the representative of the future, the ambassador from its coming neighbor, Colstrip Three and Four. Beside the plant that existed there was now the ghost of the second phase, a quadrupling of megawatts, more than a doubling of size. Three and Four cast its shadow across the towers, pipes, generators, cooling towers, and stacks of One and Two, and out into the summer pastures and hay fields of all the surrounding lands.

When Bailey drove over the hill he found it hard to conceive of a plant four times the capacity of One and Two, so in imagination the superstructure and stacks of Three and Four towered over the whole landscape, dwarfing in perspective even the Sarpy Mountains to the west. In this vast combination, Colstrip One through Four, was contained all the threat implicit in the *North Central Power Study,* the consuming of the West by these legions of machines that spoke, rather than listened, to men.

There had been a change in the mood of the county. One and Two, which had been the center of discussion, were now installed in the landscape. Soon the construction work that they made would end, and men would be back at the union hall waiting for jobs. As this inevitability began to overshadow the drive to get the first part done, Three and Four loomed over the workers, too, promising two more years of security and good wages. And merchants in Forsyth, who had reaped the payrolls created by One and Two, and had expanded with that money, began to realize that as the first boom subsided they would have higher inventories and dwindling numbers of customers. They too began to lean on the promise of Three and Four. Everyone looked to the new machine: Ranchers who had been hesitant at the beginning to commit themselves either way now saw what was there and what it had done to Colstrip and recoiled like Bailey from the prospect of more; construction workers, unaccustomed to the relative stability of a two-year-long job, longed to have another such pause in their nomadic lives; and those merchants whom the steam plant had rewarded with money and trade were filled with fear that the flow would stop.

In January 1974 a Bechtel superintendent had held out the ultimate plum: a succession of power plants, one after the other, sprouting up in a line beside One and Two. Asked by a reporter how many men he expected to be working on Three and Four, the superintendent had estimated that about eleven hundred would be needed by the summer of 1975 if the plant was approved. "But," he added, "there could be as many as eight or sixteen units to be built at Colstrip. Many ranchers," he continued, hinting that the power plant's largess would now finally be distributed to the only group left out in the cold, "may be in the real estate business sooner than they thought."

A promise by a Montana Power Company public relations executive in August that Unit Five would be the last phase of construction in Colstrip convinced neither enemies nor friends of the power plant. No one really believed it; it was simply assumed to be part of a one-hundred-thousand-dollar promotional campaign Montana Power had recently announced, which was directed at securing approval for Three and Four.

For five years Don Bailey had been fighting coal development in Rosebud County. It had started with him and Wally McRae, two young mavericks who, many of their neighbors felt, were just cutting oratorical teeth on a token issue before embarking on their political careers. It had grown into something that threatened him with two prongs: If the coal development itself didn't ruin his way of life, the time, money, and energy he spent fighting it might. But at least the issue had grown so large the neighbors were part of it now and the load was shared. And now, finally, it seemed to be working its way toward a climax. Three and Four were the test: If they were built it might just be too much, and the wisest thing might be to sell out and leave. But if they were defeated, Montana Power Company would have to go elsewhere, and the momentum of the old inexorable coal would have been broken.

In the fall of 1974 the Montana Department of Natural Resources and Conservation—a state agency—announced that it would make a massive, unprecedented effort to seek public comments on the fate of Colstrip Units Three and Four before it made recommendations to the state Board of Natural Resources and Conservation—a politically appointed panel—whether or not to approve the plant. The DNRC would hold eighteen public meetings across the state to explain the plant and listen to public testimony. A meeting would be held on December 18 in Forsyth, in the county courthouse.

This would be the big push, Bailey thought. He was ready.

37

At the Capuchin friary in Huntington, Indiana, the brothers had shaved off their beards in 1963. And after that changes came swiftly, as if the sudden appearance of chins among the brethren so stunned the order that any other modernizations seemed minor. "We rushed from the medieval age to the modern age in five years," one brother recalled. From the long robes, the silence, and the seclusion of the friary the leap into the secular world must have seemed abrupt.

It was certainly a vigorous change for that particular brother, whose name was Ted Cramer. Cramer had been a cook and a tailor in the friary; he had worn the traditional Capuchin habit with its pointed hood, and he had become accustomed to the unending routine of hours of prayer and hours of simple work. But in 1965 he had been sent to join the St. Labre Indian Mission in Montana. The mission, under the direction of a dynamic young priest, was rapidly becoming one of the most efficient fund-raising organizations in the United States, and there the rigidity and asceticism was almost entirely abandoned. "Missionary life was very unstructured," Cramer remembered. "You found your own niche, and the niche I found was flying."

So Cramer shed the sandals and cloak for the utilitarian suit of the pilot, and learned to fly the mission's twin-engine Cessna Skymaster. Once certified, he spent seven years flying Capuchin leaders, the mission's head, Father Emmett Hoffman, and school personnel to Detroit and back, from the mission's scoria landing strip. As a pilot he developed and refined a sense of detail that would later even extend into the planning of backpacking trips into the hills around Ashland, the home of the mission. "Brother Ted's very systematic," a fellow mission staff member said. "He's very organized. He likes to plan things to the minutest details. He dislikes the element of surprise."

From the days when it split from the mainstream of Franciscan doctrine in the seventeenth century, the Capuchin order was a brotherhood of beggars. Its members vowed to forsake all things worldly and to live only on charity, and to serve also as a source of charity for others in need. Father Hoffman took this simple creed and refined it into a fund-raising mechanism of unusual power. In the years when Ted Cramer was becoming active in opposing the power plant, the mission was raising about six million dollars a year from public contributions. Of this about two million dollars was used in fund raising.

The method was not complicated. From offices near the vast stone tipi-shaped chapel Hoffman had built, thousands of letters went out daily; enough to make Ashland, a town of about five hundred, rate a first-class post office. In each of these letters was a trinket, a small plastic tipi or a pen holder shaped as an animal or Indian child. The trinkets were made of premolded plastic parts shipped from New York and assembled by about one hundred Northern Cheyennes and Crows in a reservation factory. Accompanying each trinket would be a passionate, pleading letter signed by Father Hoffman.

"Dear Friend," the letter would begin, ". . . Please let me acquaint you with these destitute people, known to history as the *Race of Sorrows*, and with our work among them." One such letter contained the story of a Dolly Bighead, who was pictured in the letter hugging a cat. Dolly Bighead, the letter said, had lived at the mission for three years and was then released to her mother, "who promised to take good care of her. . . . However, at the beginning of the school term . . . it was discovered that Dolly had primary tuberculosis. . . . The meager diet of the Indian people, and the poorly ventilated one-room shack which was her home, affected the child's health." The letter also went on to tell of the plight of Morris Littlehead, who caught typhoid fever and died while on vacation at home, "another victim of poverty—lack of proper disposal methods caused the contamination.

"Our space is so limited . . . that some of the children must sleep at home (We have been hoping to build a new dormitory but our daily needs are so great and our income so limited that we have not been able to do so.) . . ."

Finally, in a P.S., Father Hoffman explained that the plastic toy in the envelope was made by Northern Cheyennes as a gift. "While I hope that you will assist us, please do keep this gift even if you cannot or prefer not to help us."

The name of the trinket with the letter, he said, was "Suzi-Starving Bear."

Other materials sent out from the mission included ads in professional football programs ("When Johnny got an A he cried"); a newsletter from Hoffman containing a form for a bequest, in case the recipient chose to remember St. Labre in his or her will ("The time has come for us to be concerned about the future"); a request for donations to cover the purchase of a piece of dental equipment which would cost $4,300, "far more than our budget can handle this year"; a brochure bearing a photograph of a wide-eyed child and the slogan "You can't dream in Busby"; and a printed copy of a letter to Santa Claus from a Northern Cheyenne boy: "My house must be hard to find caus you never come at Christmas . . . Please try to find us. We need things so bad. Could you bring a doll for my sisters? I wont mine if you dont have a toy for me. . . ."

"Will 'Santa Claus' fail him again?" asked the brochure. "Please make glad the hearts of the Children of the Race of Sorrows!"

In October 1973 the Cessna Skymaster had come due for an overhaul and the mission had decided to sell the plane. "It had always been an image difficulty," Cramer said later. "Poor Indian mission with an airplane and all that; it just didn't ring too true." So Cramer had been without a job. But about that time the mission decided that image needed some repair, and it invested in a sophisticated audio-visual center and set Cramer to work producing educational tapes and slide shows for both the students and the public. So he turned his precise mind to the mixing of sound and photographs to produce a message. "When I was into flying," he remembered, "the only thing I thought about was weather. I'm much happier doing what I'm doing now."

Back in a routine that at least kept him home, Ted Cramer started to get involved with local politics. He didn't have much luck with the Indians. "I've attempted to get involved with the Indians. But they want to do this themselves—maybe it's their turn to stand upright." But he found an open door to the Rosebud Protective Association, and the ranchers' image of underdog appealed to him.

"I don't have a vested interest in Montana," he said. "But yet I knew the people around here were being oppressed and I saw them fighting back and I kind of liked that . . . I wanted to be identified with that."

From the point of view of his communal way of life Cramer saw the capitalistic system as a monster. "I'm aware of some of the bad it does to the local people for the good of the nation." The power plant, symbolic of a greed for energy and profits at the expense of ranchers, Indians, and the land, was a natural enemy. So when he learned of the meetings being held all over the state by the Department of Natural Resources and Conservation to hear testimony on the need for Colstrip Units Three and Four, he decided to organize a contribution.

"I saw that this area needed to be spoken to, and I thought the church, the Catholic Church, should be aware of what's going on because radical changes are going to happen and unless we're aware of what's going on we are going to be spending all our time catching up. And I thought that somebody that's in a key position of leadership or power should be doing that. Somebody like Father Hoffman. But I didn't think he had the time—I didn't think he had the interest, either. So I thought maybe I could begin to do something."

Cramer once again collided with the problem that had ended his flying career: image. "Financially we probably depend on the power companies to support us," Cramer said, "and, see, it could be very touchy if we started biting the hand that fed us. If St. Labre's became radically involved, the coal companies could say, 'Hey, man, we can fix you—we'll give you some bad publicity.' And that can happen to us very easily."

But in November 1974, Ted Cramer and a Capuchin priest who was also interested in coal development drafted a letter to be read at a meeting in Ashland and at Forsyth. But it was signed not in the name of St. Labre, but in the name of the Capuchin community at Ashland. "That's how it was defined," Cramer said. "But some people don't see a difference."

By the beginning of December the letter was finished. And Ted Cramer, although he hated public speaking, was ready to deliver it at Forsyth. The letter was more than just a small victory for Cramer in his drive to get the order to fight openly against the power plant; it was signed by nine priests and brothers, including Father Hoffman, and although it said at the start that it just represented the Capuchin community, it was typed under the letterhead of the St. Labre Indian School. And there was power.

Cramer waited nervously for the day to come. He was ready.

38

Two studies on the impact of the coal rush differ: Dr. Paul E. Polzin, of the Bureau of Business and Economic Research at the University of Montana, Missoula, says coal development will add, at most, about 20,000 additional people to southeastern Montana by 1985. . . . The Northern Great Plains study predicts that . . . by the year 2000, the population . . . of Forsyth will be 6,000; Colstrip 13,000 and Lame Deer 6,000. The Northern Great Plains study concludes that these towns are likely to experience the boom-town syndrome, meaning "a whole family of mental health symptoms and problems, such as rapid influx of people, families crowded into mobile home camps, increasing crime, alcoholism and suicide rates.

"In short," the federal report stated, "the quality of life of persons, both newcomers and residents of the area, is degraded."

—*Forsyth Independent*
November 28, 1974

Mike Hayworth woke up at 3:30 in the morning on December 18, 1974. He lay awake in the darkness of his home in Colstrip listening to the silence and the breathing of his wife. The power plant sometimes made strange noises in the night—clanks and shufflings and slow rumbling sounds as the night shift continued what the day shift had begun and the machine got closer to life, but few of them reached all the way over here on the other side of town. So Hayworth had plenty of quiet in which to listen to himself and to the approach of the conviction, rushing at him with the nervous clattering energy of a train, that tonight he was going to have to speak.

Mike Hayworth was twenty-eight. He had been raised on a family farm near Geraldine, Montana, up near Great Falls. He had come to Colstrip in 1970 to teach in the high school, but in the spring of 1974 he had quit that job to go to work at the power plant as a plant chemist. "It was the type of work I had intended to go into when I first went to college," he remembered, "and the job wasn't going to be terminated when they stopped construction. The job had a very bright future."

Mike Hayworth had no ambivalence about the power plant, although his older brother, Pat, lived down on the Rosebud and worked for Don Bailey, and was opposed to coal development. The plant had been good to Mike. "Given my background as a farm boy," he said later, "I'm quite proud that I was able to find something right here in Montana as exciting as this . . . as rewarding. If Three and Four are built there's a tremendous opportunity to move up." And he was a man prone to intense convictions. So when he woke up at 3:30 A.M. he knew he had to speak. For several days he had been reading the testimony that ranchers and other Colstrip residents had been giving at the series of public meetings that were being held all over the state, and now he knew it was his turn. So he got out of bed and walked quietly into the kitchen. He turned on the light and spread copies of several newspapers out on the table. Then he sat down, prayed quietly for a moment, and began to write. If the thought that perhaps his brother might be doing the same thing ever crossed his mind, it didn't linger. Pat was a shy, somewhat taciturn man; *he'd* never get up and speak.

Mike Hayworth had first become publicly involved in the controversy over the power plant two weeks before. On November 29 he had read a letter from a Montana soldier stationed in Germany·that had been printed in the *Billings Gazette.* The letter had praised the qualities of Montana that the writer loved, "clean air, water, and wide open spaces," then went on: "Yet we let companies like Montana Power move in, remove our topsoil and pollute our air and water."

Hayworth had been irritated, so he had written a reply. In it he had argued that the plant and mines damaged the Montana environment hardly more than overgrazing, that the reclamation on the mined land worked, and that the power plant would ensure the soldier that he would have a job available for him when he returned from overseas. And he added one piece of philosophy:

"In considering the pro- and anti-Colstrip problem, state, national, and world problems, I can find one common denominator. Greed is the common word for it. Greed for money, power, influence, etc.

"I see greed in myself, the farmers and ranchers I've worked for, organizations I've belonged to, corporations, government, etc. Actually a complete explanation of the problem can be found in the Bible.

"We all have the same hang-up and, yes, the solution is available to all who will accept it. (See Romans 3:23, 6:23 and John 3:16 for starters.)"

Among the clippings on Mike Hayworth's desk was a report in the *Forsyth Independent* on a meeting held in Bozeman a few days before. One of the speakers at the meeting there had been Duke McRae, who had mentioned the problems caused by crowding at the Colstrip schools and who had suggested that since Montana Power Company created 100 percent of the impact, the company should provide some payments to ease it. Mike Hayworth, interpreting this to be an attack on the school system in which he had taught until so recently, was irritated. But he thought he knew why McRae was so outspoken. He thought he had him figured.

A few years before, in a conflict between several teachers and the administration in the Colstrip school, Patricia McRae was one of a group of teachers who had lost their jobs. She had gone to work at St. Labre. Although the ranching community had long ago absorbed and tried to diffuse the anger and division that had accompanied the incident, Hayworth remembered it. It seemed logical to him that McRae harbored hidden grudges. So when he sat down to write that was in the top of his mind. He'd let the public know what hidden motivations were involved here. He'd strip the sham from the ugliness. He'd let the people know what the *truth* was. Too bad if the truth sometimes hurt.

Hayworth wrote for hours, went to work, came home, and wrote again. And finally he drove down to Forsyth with a typed speech several pages long. He was early. He parked the car on the street and walked up the snowy walk between bare trees to enter on the lower level. The halls were quiet. The door of the sheriff's office was open and there was the sound of a police radio squawking routinely, but the other offices were locked and dark. There was only one other person in the hall.

It was his brother Pat.

"Hi, Pat," he said.

Pat Hayworth smiled. He was a quiet man, with a snub nose and a friendly face.

"Hi, Mike." Mike could see, in his brother's hand, a single sheet of folded paper. He held his own wad of papers a bit self-consciously at his side.

"You going to give a speech, too?" Mike asked.

"Yeah, I guess." Pat was slightly sheepish. "Got to say what I feel."

There was an awkwardness. Someone neither of them knew came in and nodded at them as he passed and went up the stairs.

"Well," Pat said. "Kind of fun. Two brothers, one on one side, one on the other side. I believe it's just like the Civil War."

The two men laughed. And they went up the stairs to the courtroom, which waited, still almost empty, for the meeting to come and fill it with anger, with people, and with noise. But they didn't go in together.

Part 2

39

Mike Hayworth was one of the first. When he came through the big oak doors he glanced involuntarily up at the flamboyant mural of Moses carrying the commandments and then went up front to sit down. The seventy-five chairs back of the rail between the spectators and the enclosure up front were hard wooden theater seats; the jury seats were padded. Hayworth walked up, took a jury seat, and sat down to wait.

Pat Hayworth came in moments later and sat down in the middle of the spectator area, folding down one of the hard seats and looking around at the surprising number of people who were already there. Going to be crowded, he thought. He turned the folded sheet of paper around in his hand, opened it, looked at the six short paragraphs. They were typed on a machine that had lost all sense of the horizontal, so the words twitched their way across the page. It's going to be so crowded, Hayworth thought, I'll probably get out of here without having to talk. He settled back in the chair and put the folded paper away in his shirt pocket.

Wally McRae was also early. He had come with another rancher, but when he walked into the room, wary as a coyote, his hair brushed back and his face looking leaner than ever because he wasn't wearing his scarf, he went to sit apart from his friends. At the meeting held the previous day, in Hardin, McRae had talked to Mike Moon, the moderator, after the event. "Can't have that meeting tomorrow in the courtroom," he had said. "You'll have a riot. It's too small. Put a sign on the door, move it to the library."

But Moon had been adamant. "You've got to be so careful. We've announced it in all the papers for the courtroom, and now that's where it's got to be." So McRae was nervous, not because he was going to speak, but because of what might happen.

151

Duke, Pat, and Sheila McRae came in together. They sat near the middle of the left side of the room, near Don Bailey. Bailey was leaning back in his chair, his feet stretched out in front of him, his arms folded, his jaw muscles working. Sheila glanced at him, at the mural, and around the room. The place looked segregated. The left side, where she was sitting, seemed to be dominated by ranchers. Bailey, Bill Gillin, and there was Barbara Wallace with Charlie and her mother and father. Barbara looked serene, but then she had had experience in front of crowds. On the other side of the room were most of the people she knew from Colstrip. She wondered if Mike's parents were there. She looked down at her hands. They were still. She suddenly noticed the heat. She looked out toward a window, and gave the room her profile.

Marie Sanchez came in alone and sat in the back. She could see only one other Indian in the room: Myron Brien, who was sitting calmly near the front, neither fear nor excitement in his eyes. She wondered when the Northern Cheyennes would finally learn what was important.

Clifford Powell got in just in time to get a seat in the far back. He settled in, folded his arms, and nodded at another pipefitter who was sitting down front. The other man waved his clenched fist. Powell showed his teeth. Lot of people here. Most everybody except for the men on the swing shift, whose first duty was to the power plant.

Mike Moon came early, but not early enough. By the time he got to the room, half an hour before the meeting was scheduled to start, it was already nearly full. Only some of the seats down front were empty, and the area in front of the little wall was unoccupied. There was already some congestion in the back. "Why don't you fill in down in here," Moon said, putting the tape recorder down on a table. "We're going to need all the room we can get."

There was a general shuffling. Wally McRae came forward through the little gate and sat down against the wall under a table, looking across at the jury seats, which now seemed to be filled with Montana Power executives. Above McRae on the wall was a black-and-white copy of a portrait of George Washington, who looked as if he had just been caught in his attic reading a cowboy novel. Other people came up through the gate and settled in on the floor with their backs to the side walls and the little fence; McRae was joined under the table by a younger man with a sharp-edged aggressive face, longer hair, and a sardonic air. They introduced themselves: "Wally McRae." "Vic Jungers."

The rearrangement opened a few seats in the back. Into one of them slipped Ted Cramer, trying not to look self-conscious in his black suit and tight white collar whose restriction was so unfamiliar. But he was not displeased that the costume was noticed.

When Lois Olmstead came in with her husband all the seats were taken, and there were men standing in the back. She had worn a blue skirt and a red plaid jacket to the meeting, and now, looking around the room, she realized she was probably the only woman not wearing slacks. She was proud of her femininity; when a man offered her his seat she smiled and thanked him and sat down, and didn't notice if he was a rancher or a construction worker or a man in supervision. She looked out across the crowd and saw Nick Golder way over against a wall, looking cold and hard, and she thought: If only we could just talk together, rather than all this big rigamarole.

Mike Moon, who was trying to get the tape recorder to work, began to hear commotion in the hall. It was now past 7:30, the time set for the meeting to start, and the room would hold no more. Even the area behind the judge's chair was now full. Yet there was a muffled sound of crowds outside in the hall, and out of this came an occasional disorganized chant, "We want in, we want in." Moon began to long for the library. He even suggested the possibility to the people in the first few rows, and it was discussed for a moment, but Mike Hayworth got up from his chair in the jury box and said, "Sure, you'll move it to the library and all the people outside will get the seats and we'll be outside. I don't like that idea." So the meeting stayed where it was.

But the rumor floated back through the room and out the door, and some of those waiting in the hall heard it as a fact and went back downstairs, out the door, and a couple of blocks west to the library. One of them was Bill Parker. He helped set up chairs and tables in the long, low room in the library's basement, but when no one came he drifted back over to the courtroom and had to stand outside the doors with his fistful of testimony, waiting for his chance.

And Martin White, too, was out in the hall. But he had seen Don Bailey before Bailey went in, so when the three- by five-inch cards came around for the people who wanted to speak to sign, Bailey took two, and he signed his own name to the first and White's name to the second. Bailey had notes, but White didn't have any idea what he himself was going to say. He'd find something, he knew, when Moon called his name.

McRae and Bailey signed the cards almost out of habit. Duke

McRae signed his proper name with a sigh and passed the stack of cards to his wife and daughter. Patricia McRae signed vigorously; Sheila signed quickly, then passed the cards on with a little involuntary grimace. She was watching the other side of the room. Barbara and Charlie Wallace signed the cards as a precaution, not knowing yet whether they wanted to speak. Tom Wimer signed a little reluctantly—he didn't like to speak unless he had to. Ted Cramer signed without hesitation. Lois Olmstead signed. Clifford Powell signed. Myron Brien signed. Mike Hayworth signed. Even Bill Parker, out in the hall, managed to get hold of a card, write his name, and send it back into the room like a letter begging for an audience with a king. And Pat Hayworth signed, carrying through the intention he had built the last few days, although he was almost sure that all the eloquent ones would be called and he would be able to escape unheard.

But Vic Jungers, still the skeptical observer, still removed enough from the conflict to be uncertain even of which side he was on, didn't sign.

A coal train passed, howling at the level crossing, clattering on the tracks, and fading away into the distance, east. Mike Moon presented a brief slide show on the power plant's environmental impact. The chanting in the hall subsided as the Rosebud County Sheriff's Department officers arrived. Then the lights came back on and Mike Moon shuffled the names. A small hum of tension went through the room, like the nearly imperceptible sound of a new bowstring tightening down. Mike Moon looked at the first card. He said:

"Patrick Hayworth."

40

Following is an abridged transcript of the testimony of eighteen persons at the meeting called by the Department of Natural Resources and Conservation to gather input on Colstrip generating Units Three and Four held on December 18, 1974, in the courtroom of the Forsyth County Courthouse from 7:30 P.M. until midnight. Speeches are reported in the order in which they were made. Parenthetical notes are provided by the author.

Patrick Hayworth

["I was scared," Pat Hayworth remembered. "I mean, you know, well, I'm sure no orator, you know. And it wasn't that I didn't believe in what I was speaking on or speaking against, you know. It was just getting up in front of all them people." "Pat is absolutely one of the most quiet, meek human beings I've ever been around," Wally McRae said. "There were two people who really had a lot of guts to get up there. Pat was one of them."]

I am Pat Hayworth. I am employed on a ranch on the Rosebud Creek. I am speaking on behalf of myself and my family, and anyone else who realizes the importance of agriculture to our community and country.

It has been stated at previous hearings that good job opportunities have never been available to those who would choose ranching as a livelihood.

I have grown up and lived on ranches all of my life, and I believe I have contributed as much to a community as has anyone employed in any other type of employment.

I believe that anyone who wishes to endeavor, and prove

155

their value to an outfit, can find employment in agriculture that will afford as good a living standard, and probably more personal satisfaction, as any other type of employment.

My salary plus benefits, today, is $12,200 a year plus a—ah—psychic wage that might nearly equal that, to people who have had to live in the scourge of heavily industrialized areas.

Because I believe the construction of Units Three and Four at Colstrip may set a precedent that will lead to the eventual destruction of agriculture in this region, and the—ah—termination of my job, I oppose their construction. [*Applause.*]

Patricia McRae

[*"My mother writes like a teacher," Sheila McRae said once, in appreciation. "Even her letters to me are—they don't sound like a mother's letters to a daughter. They sound like a book or something." But Patricia McRae's cool, steady delivery, a straightforward reading of a typed manuscript, muffled the power of her words, and while she spoke there were snickers from the construction workers on the right side of the room.*]

I'm Patricia McRae. I'm a high school English teacher at St. Labre Indian School. Deep inside I'm a poet.

As I drive to work each morning across the Northern Cheyenne Reservation, which is directly in line of flow of the wind from Colstrip, the awesome majesty of the hills feeds my spirit and I feel very close to nature and our neighbors. The scene I see is never the same. The sun, the rain, mist, snow, paint a different picture every morning. But whatever the view, it is always inspiring.

I am not naive to believe that the power company is as humanitarian as it would like us to believe. . . . Montana Power can brush aside a man without a second thought while proclaiming loudly how much they're helping the people of Montana. I have no quarrel with any individual. We are all pawns, to be used by the corporate giant to increase their profits, and when our usefulness is at an end, to be dropped by the wayside. Their stockholders see figures on the annual report, not majestic hills or sandlots. This country took millions of years to form, to create the beauty; and corporate man is going to snatch it away in a snap of his giant shovel jaws. We do not know what the plants will do. We can all make guesses, and the only sure conclusion is that there will be a deterioration in the quality of our life, our water, our air, our land. No one in his wildest predictions sees an improvement in this area. The only disagreement is the degree of destruction.

Poetry, according to Robert Frost, is only a small cog in the machinery of life, but it's a vital cog. Man doth not live by bread alone. The beauty of this area is poetry for many people. . . . Tonight I am pleading for the peace of the hilltop; for the song of the meadowlark on a clear, dry morning, for the things we all need to restore our sanity . . . the things which are being pushed further and further from our grasp.

No lightbulb can duplicate a late afternoon sky, and its sinking sun, with magnificent gold, purple and yellow green. No air conditioner can refresh us with the tangy breeze of sagebrush after a summer storm. Unlike the Wizard of Oz, I am unable to give Montana Power a heart. . . . Therefore I oppose Colstrip Plants Three and Four. After the coal is gone, after the plants are constructed . . . No company on earth can re-create the beauty of this area for the children who are to come. [*Applause.*]

Clifford Powell

[*Powell held the microphone right next to his mouth. When he spoke the recording needle of the tape recorder leaped against the stop in an agony of overload. He was out of breath from the brisk walk to the front of the room; his first sentences came in bursts.*]

I'm Cliff Powell and I'm from Billings. And my speech is not as elegant—as the lady—just made. I heard a comment about a—meadowlark singing but—you ever hear some hungry kids bawl?

I belong to the plumbers and pipefitters local at the present time. And I work in Colstrip. And I support Units Three and Four. There are more than just farmers and ranchers in Montana. There are also workin' people outside of agriculture. And they need to make a livin' too.

I'm a fourth-generation Montanan, and I feel that I got the right to live and make a livin' here just like the farmer and rancher does. The farm and ranch pay doesn't make a livin' for me. We need industry and growth. And I think that both factions can live and work together if we try. I'm not for running rough-shod over anybody's land. I'm not for big smokestacks billowing black columns of smoke polluting the air. But Montana Power's meeting the full standards and requirements set by the law. We're part of that; we put the people there that made these laws. All we have to do is make sure that the laws are enforced.

Right now we need energy so bad in all its forms, and a small few are battling it. It's more than just Montana, we can't put a fence around ourself, there's a whole world around us. We export beef, pork, lamb, wheat, oats, barley, sugarbeets, peas, and many more agricultural products. And yet we don't want to export electricity or anything else.

Today in the news, NBC, they said that 75 percent of the wheat of Montana goes to Asia, so why the hell don't we raise wheat in Asia and sell it here? That's cheaper too. [*Applause, mingled with shouting. A voice: "All* right!"]

We're much more capable of a diversified economy than just agriculture. All we got to do is get together and get with it and we can all work together and make it go. There's no use fighting—one another. [*Applause. "It was like a melodrama all night," Wally McRae remembered. "There was booing and hissing and clapping and elbowing and a lot of facial expressions, trying to bolster up and support your side. Everybody knew it was a confrontation and nobody wanted to overdo it, but they were trying to hit the happy medium on supporting their guy and their team and waving their pennant. But you kind of looked over your shoulder when you waved your pennant, you know, because you weren't sure but what somebody was going to hit you with a bottle of Jim Beam from the top row in the bleachers."*]

Sheila McRae

[*"In a way I felt like—just because my name was McRae I should say I don't think the steam plant should go through," Sheila McRae remembered much later. "But I had a lot of friends who worked there and I didn't want them to leave, and—it seemed if it didn't go through, they'd leave. So I really . . . I think I did it because Dad thought it would help if I did. I don't know."*]

My name is Sheila McRae. I'm seventeen years old and a senior at Colstrip High School. I am not as knowledgeable as most of you about the steam plant, and I don't know any astonishing figures about them except the ones I see in the news. I have seen Colstrip change from the small, friendly, caring, healthy town to what it is now: an ugly, unorganized, fast-growing town. It used to be safe to decorate the Christmas tree in the park. Now the lights on it don't stay there for more than a few days. The people are different now than before. Drugs are a problem in high school. I had never seen drugs before the power plant started. [*A solitary titter of female laughter from the crowd.*]

There is nothing else to do for lots of people. No one wants to get involved in finding something else to do. Why should they? They're only here for the job, and maybe the next place they go will be better. Or so they hope. [*Applause, followed by a small spate of coughing, as if people had been holding their breaths.*]

Tom Wimer

[*"Sometimes we had a hard time rallying other people to help us," Don Bailey remembered. "They'd say 'You guys are doin' great and keep it up.' But you'd go and ask them to go to a hearing and, 'Well, I just don't have the time. Can't make it.' We really had to orchestrate those people, get them out to participate in the hearings." "I don't think I had any real strong feelings," Tom Wimer said later. "I just felt that I had certain ideas that I wanted to express real brief."*]

I'm Tom Wimer and I live in Colstrip. I'm a rancher. I think due to the social, environmental, and economic impact on this area, due to plants number One and Two, there should be no consideration at this time of building any more plants. Until One and Two are stabilized and operating and have been evaluated. Also I think if there is a power need in the Northwest, we should ship the power there where they have the water and the population to take care of it. [*Applause.*]

Mike Hayworth

[*"I'll have to admit," Mike Hayworth recalled, "I was nervous, jittery. I was just writing little notes down as each one of the previous ones testified. And you really didn't know when you were going to testify. I was nervous because here I'm going to say things that probably don't exactly sit well with the people I'm saying them to."*]

I apologize right now if I take more than five minutes of your time, but I do want you to realize that I got up at 3:00 A.M. . . . and have been working till six on my statement, so please bear with me.

I can spend ten minutes countering some of the things that have been said here tonight. . . . Say, did I forget to tell you?—I'm plant chemist right now at Colstrip One and Two, my address is Colstrip.

About the lights on the Christmas tree. Ask Obert Rye— Obert used to have to keep an eye on the lights five years ago—I talked to him—I was there. . . .

Okay, well, let me get on with my statement here. I moved to Colstrip in 1970 when I was hired as a physical science and math teacher in the Colstrip school. I grew up on a farm in the central part of Montana near the small town of Geraldine. I received my teaching degree at WMC in Dillon. When I decided to accept a teaching position at Colstrip very few people I talked to at that time had heard of Colstrip. [*He spoke very rapidly, flinging the sentences out one after the other.*] When my wife, oldest son and I moved to Colstrip in 1970 there were several vacant houses. So we've seen Colstrip as it grew the past four and a half years. . . . It is good that the [*Department of Natural Resources and Conservation*] came to Forsyth for this meeting because the closer people live to Colstrip the more informed they are and the less apt people are to make absurd statements such as were made at the Bozeman meeting, concerning the school.

The person making the derogatory remarks in Bozeman concerning the school seems to have some sour grapes in his system concerning personnel changes that took place in the Colstrip school system five years ago. Let's clear the air of these personal grudges and start acting out of genuine concern for the educational needs of our children. [*There was a rustling among the ranchers. Later a Colstrip worker who had been sitting on that side of the room reported to Hayworth that a member of one ranching family had turned to another and said: "That son of a bitch!"*] The present administration at the Colstrip school is very conscientious about such things as state standards, accreditation, and individual needs of the students. A common phrase I hear is that there has been a deterioration in the educational standards in Colstrip. I taught there, and I'm aware of some changes. Colstrip had a reputation of being a college prep school. Presently, in addition to college prep, the school is introducing career education and vocational education, classes that better meet the needs of the majority of the students. [*Hayworth stopped to turn a page. The effect was as if a charging rhinoceros had suddenly paused to tie a shoelace. At this moment Barbara Wallace decided she would speak; when her card came up she would not pass.*]

The present condition in the school or any personal complaint anyone has about the school has no place in considering the impact, final outcome of, building generating plants. Let's keep the schools out of this. . . .

I certainly hope I didn't give someone a bad image of ranchers, because I consider them my good friends. I've worked and visited with them; they've allowed me to hunt on their land. They need to make a living too, and I give them lots of credit for standing up and making their viewpoint known. I certainly hope you've listened to my viewpoint and give it due consideration. [*Applause.*]

Brother Ted Cramer

[*"I never wear a Roman collar," Ted Cramer remembered. "But I wore a Roman collar when I went there because I knew the impact it would have— I wanted people to know who I was and who I was representing. It may have appeared kind of phony, but I think that was not my intent. I wanted to give—visually I wanted to make a bigger impact." His voice as he spoke to the crowd was cool and without passion; after the assault of Hayworth's voice it was almost soothing. But the sheet of paper he held shook slightly in his hand.*]

My name is Brother Ted Cramer; I'm from St. Labre's Indian School. I'd like to make this statement in the name of the undersigned Capuchin fathers and brothers residing at the St. Labre Mission, Ashland, Montana, and surrounding mission stations.

As ministers in this area we are concerned regarding the construction of Plants Three and Four at Colstrip, Montana. I will express these concerns through the use of a few questions:

What will be the precise impact of Units Three and Four on the lives of the people in this area? Will they help provide a better way of life?

And from the previous question, what effect will Units Three and Four have on the spiritual life of the people?

Does the principle of what is good for ten—or even a hundred—justify disregard for the rights of an individual?

How can the needs of the people in cities be met without destroying the quality or way of life for the people in this area?

What effect will Units Three and Four have on the culture of both the Indian and ranching communities? . . .

We must honestly say that we do not know the answers to some of the above questions. At present Plants One and Two are nearing completion. Soon they will be in operation. At that time we will be able to experience the effect these plants will have on our environment and on our people. We propose that a moratorium be placed on Plants Three and Four until we have gained this valuable information. To move ahead before that time would be to move in darkness. . . . Thank you. [*Applause. Now Martin White had found what he wanted to say.*]

Wallace McRae

[*McRae's opening sentences were brief and crisp, emphasized by long pauses. The rustling, giggling, and elbowing in the audience ended, as it does at any game where a performance is unusually compelling. Three men from the crowd at the back who had gone out to the hall to smoke drifted back into the room. But Wally McRae remembered: "Not one of my best."*]

My name is Wally McRae. My address is Forsyth. I'm a member of the Rosebud Protective Association, and chairman of the Northern Plains Resource Council. And I'm speaking tonight on my own behalf. . . .

One of the things that bothers me is this: This will be the fifth meeting I've been to, and so far there has been no representative of any of the applicant companies other than Montana Power. . . . It seems odd to me that Montana Power is the only applicant company that is being visible. And I don't know. There is a kind of a code of the West type of thing, when you're going to a new community and you're going to have some new neighbors, one of the first things you do if you have a sense of propriety is to go around and meet your new neighbors. And find out what they expect of you and what you expect of them. And I think that if the other utility companies are serious about coming to Colstrip and being neighbors, that they would be here. [*Long pause.*] And they aren't.

Now this tells one of two things to me. Either they don't care about being good neighbors, when they get here. Or the second possibility—maybe they aren't figuring on coming here at all and I hope the second is the case.

One of the Montana Power people said last night that . . . maybe a small number of people would probably suffer, but he said we have to put this in perspective—I'm paraphrasing, but I think this is the main crux of what he said. He said we have to keep in mind the goal of the greatest good for the greatest number of people. [*Long pause.*]

And you know, this is the psychology, this is the philosophy, of the socialistic and the totalitarian states. This type of psychology—the greatest good for the greatest number of people—is what happened here just a little over a hundred years ago. When the Indian was a problem for economic development, for resource development, and the greatest good for the greatest number of people dictated that the red menace be eliminated. And it was done.

And I hope that this isn't a pervading psychology of our new applicant neighbors. [*Long pause.*]

And Patrick Henry said one time "Give me liberty or give me death." Well, a lot of people are saying the same thing now, only they've forgotten the last part of it. They've forgotten about liberty and death and all they say is "Give me."

I think that we are talking about more than a decision about two more power plants for Colstrip. . . . There was a French philosopher named Raynal who addressed this very well. And he said there is an infinity of political errors, which once adopted become principle. And I don't think that we're talking about just two plants. I think we're talking about solving the problems in Green

Bay and Sheboygan, also; and I hope that we don't make this political error so that it will become principle. And I am opposed to Colstrip Three and Four. [*Applause.*]

Lois Olmstead

[*"I had no plans to get up and say something at that hearing," Lois Olmstead remembered. "I always had the feeling that if I said anything they would say well, naturally you're going to take their side because your husband works for the power company and you don't—you can't have any opinions of your own. I remember when I had made up my mind, when I finally decided that somebody has got to say something about the people here, you know. I can see that long aisle; I can remember: 'Boy, here we go.' "*]

I'm Lois Olmstead. I'm a Colstrip homemaker, from a family of ranchers in Montana. I'd like to speak as a mother because not too many people have gotten up and spoken about a mother with children in the Colstrip schools. We are very much in favor of the schools at Colstrip; we think we're getting a top education.... But my kids are [*also*] getting an education in things about the world. They are getting to meet kids from all over the United States. And when they go out from high school, they're going to go out with a lot broader horizon than I did.

And then I would like to mention to the Department of Natural Resources that I question your survey.... I would like to state an example of how I feel the questionnaires are biased. The first time I was interviewed ... I was watching what she was writing down as I was giving my statement. In particular she asked me if we did a lot of walking in Colstrip. I said well, not as much as I'd like to. I said, when we first moved down to Colstrip I used to walk over to the phone booth because we didn't have a phone yet, and so I did more walking then than now, and I looked down and she'd written "Lack of telephone caused problems in Colstrip." [*Laughter.*]

The last thing I'd like to say is that we have three boys in our family, and a lot of ranches in Montana aren't like the ranches in Rosebud County, where they can support the next generation and then the next generation. A lot of ranches in Montana are one-family ranches and you can't stay on your dad's ranch, so you have to go out and get another job, and you'd like to stay in Montana. So for the welfare of my family, because I'd like to have energy and electricity for the next few years, and because I'd like my boys to be able to stay in Montana, I'm for Three and Four. [*Applause.*]

Evan McRae

[*"I'm not a public speaker," Duke McRae said once. "Absolutely not. It's a pretty tough job for me. I spoke at six or seven of those meetings. As far as I'm concerned, if they ever have another one of those decisions, these meetings—they should never be done again. It becomes just all personalities."*]

My name is Evan McRae. My family and I live on a ranch nine miles south of Colstrip.

I happen to know . . . that the gentleman who made the derogatory remark was not in Bozeman, and I was quoted out of context. I was speaking about taxes, and the statement that came out in the Forsyth paper—and I apologize if I have said anything that is detrimental to the schools in Colstrip because I was on the board for nine years and I have a lot at stake there too. However, I don't intend to step, to stoop to the lowness of these people.

I speak tonight as a minority individual fighting for my way of life. The proposed Colstrip project is an increment with many more to follow in industrialization of a predominantly agricultural region. The social and economic impacts on this region are enormous. Although it is sometimes a matter of personal preference and values, I feel my way of life, the lifestyle of agriculture that we know will be lost forever. My ranch, which can produce crops and livestock and taxes to provide a living, has been doing so for ninety-one years. A power plant has a rather short life. Likewise the coal deposit once mined is depleted forever. Continued taxes and jobs past the life of the plant will be dependent upon other industry or agriculture if there is any left after industrialization.

So I feel the choice we make for the future of our state must be made with the full realization of the consequences.

In closing I'm reminded of a quote by Adlai Stevenson: "If

this comes to pass, I'm too old to cry, and it hurts too much not to." So I'm opposed to Plants Three and Four. [*Applause.*]

Barbara Wallace

[*"I wasn't going to speak particularly when I went in,"* Barbara Wallace remembered later. *"We just wanted to listen, and as I heard some of the opposition speak I just felt like I had something to say, so I got up and spoke."*]

My name is Barbara Wallace. I live on a ranch near Colstrip on the Rosebud Creek, and I'm also a secretary at the steam plant. I have two things I'd like to say tonight. First of all, in response to something Mike Hayworth brought out, I don't know a lot of scientific facts, but I do know what it is to feel compassion in your heart for a place you've known as home for all your life. Some of you here tonight probably won't understand what I'm saying, because you haven't known a place that long.

[*"There was absolute silence when she was speaking,"* Wally McRae remembered. *"Because it was so intense and it was so difficult for her. She had tears in her eyes and it was a combination of frustration and fear and anger. 'Cause she thought: 'Here goes my job but by God, I'll tell you one thing, folks, I'm goin' in style!'"*]

Secondly, I'd like to say that at some of the hearings it has been brought to my attention that there's been some slanderous statements made toward ranchers and ranch wives and that there have been people paid to go to these hearings and say how much they like their jobs and how much they're getting paid.

I want to point out at this time that it seems very unimportant to me to even say these things at these hearings. As I said, I work at the power plant as a secretary. It's a construction job, therefore my job is temporary. There are a lot of people here tonight that work at the steam plant. Theirs are construction jobs, also temporary. When the project is built, their jobs terminate, and they move on to the next. Therefore, I think it's quite irrelevant if someone makes fifteen thousand dollars, likes his job at the steam plant or whatever. We need to discuss the important aspects of this subject. . . . We need to consider the aspects that will affect the people that will remain in our community after the plants are finished with their construction. Thank you. [*Applause.*]

Marie Sanchez

[*"I felt I didn't want to get too radical because of the setting, you know,"*
Marie Sanchez remembered. "First of all, it was in a non-Indian setting. I,
ah, changed what I was going to say. . . . " "Occasionally someone would
talk and I didn't know how to handle it," Mike Moon said later. "And she
was one."]

My name is Marie Sanchez. I am a Northern Cheyenne Indian from the Northern Cheyenne Reservation from Lame Deer,
Montana. And I represent the American Indian Movement. And
my purpose here is not necessarily for the—ah—proposal for
Three and Four. As was stated um, right now, ah. When I listen to
these people—that spoke here. Everything was in economy and.
Ah. Ah. People having to work and everything. They never think
of a people that. Can. Exist in the Rosebud community. Such as
the Northern Cheyenne people. We exist, sir, and we do have a
culture. We have religion. We have— [*Her testimony now broke down*
into brief phrases separated by long pauses, some as long as thirty seconds.]
We have politics. We have—We do exist as a people. And it
seems   like the state level never looks to us   As a people. We have   We have   water rights   We
have   We   have   sovereignty
rights.   And yet—people are such people as environmentalists trying to help us   we   We
have   They   never   try   to   think   of
                             Of   our   religion.   [*Cough*]
                                   My people try to fight for
sovereignty   They   try
to   [*Cough.*] 'scuse me.
What I'm trying to say is that we are still members of the Rosebud community   people, and we should try to be more
sensitive, ah—   [*A whispered voice: "Are you for it or against*
it? The power plant. Just tell them if you're for or against it."]
As a Cheyenne and as a member of the Rosebud Commu-
nity, I am against Three and Four, and I would do anything in
my power to stop Plants Three and Four. [*One person started clap-*
ping and for a moment rang alone like brutal sarcasm as Sanchez continued
to stand gripping the microphone. Then more hands picked up the cue and let
her walk away to applause.]

Charlie Wallace

[*"Charlie had nothing against people working at Colstrip," Barbara Wallace remembered. "He probably understood as much as anybody from being from a large city, that everybody has got to make a living. But he really hated to see a country tore up." Charlie Wallace didn't go up to the microphone; he just stood in place and spoke. On the tape, later, his voice sounded very faint and distant.*]

I don't have much to say, but I'd like to say it. I live on a ranch on Rosebud Creek, and I think if they need this power here, I think they should use Units One and Two as a precedent to see what sort of a social and environmental impact they have on this area before they go on with Units Three and Four. I stand against Three and Four. [*Applause.*]

Martin White

[*"I wasn't always that proud of what I had to say at these meetings," Martin White recalled. "Because a lot of times all the people I knew had come to say something for our side weren't moving and so I walked up and said something to try and break the ice. But the main reason I said that in Forsyth, I was goddamn irritated with that man that stood up there in his black attire as a Catholic priest."*]

[*Voice in background: "Git 'em, Martin."*] Gee, I have to thank Don Bailey for making it up here. I have been outside in the hall and I saw Don go by and I guess he put my name up here too, so thanks, Don.

[*From the audience, Don Bailey: "I just said you'd be nice, now."*]

Oh, okay. I've been listening—Oh, my name is Martin White, and I live at Colstrip and I'm here tonight as a private citizen.

I have heard a little talk here from a religious group from over at Ashland and I would like to ask that group: I know that their religion encourages propagation of our race and yet it seemed like they were a little against supplying these people, and I would like to ask them what they would like to do with these people they are encouraging their constituents to, to propagate.

There's something that we should note, and that's that there has been a steady decline in the number of ranches in Rosebud County. I don't know if any of you are aware of the fact that in 1930 there were 940 farms in the county, and in 1974 there are 320. If we continue to decrease the number of ranches in Rosebud County at this rate by 1990 there will be 96 here. And I would like to ask these ranchers who are here, what ones want to leave; which ones want to give up their place, and when they do give them up, which apparently they're going to have to, what are they going to do? You people here seem to think that you don't want 'em around; if they're going to work at something other than ranching, send 'em someplace else.

We've got a lot of these folks working here at Colstrip. In 1972 we ran a little survey on our mine out there; we had 159 employees. Eighty-seven of those had been born and raised within thirty miles of our mine. And out of that total only one was out of state other than ten that came from Cowley, Wyoming. . . .

Last night I heard a fellow say that maybe it's not fair to have the majority rule because once in a while you'll have someone who is right accused of a wrong unfairly just because he's in the minority. But I think if we look at history, the opposite has been true more frequently. Hitler and Mussolini and people like this, where one man had a tremendous amount of power, has caused a lot more trouble than where the majority of the people ruled. Which in our country we seem to think is a good method. I hope the Department of Natural Resources, when they are considering this, will think of the fact that it is one person, one vote in this country still. . . . Thank you. [*Applause.*]

Myron Brien

[*Early in his tour of Montana, speaking at the meetings, Myron Brien had sounded a bit like brochures for St. Labre Indian Mission. "For the first time I make enough to feed and clothe my family," he told an audience in Bozeman. "I can buy new boots when I need them. I don't have to wear the old ones wrapped with friction tape." But by the time he got to Forsyth he had refined his style.*]

My name is Myron Brien and I live and work at Colstrip. I am an enrolled member of the Northern Cheyenne tribe. . . .

About everything that I had in mind to say has already been said, but I would like to add that since going to work at Colstrip I've made life for myself and my family a little more comfortable. And I've been able to pay more taxes to this county, the state, and federal government as a result of making more money. That seems to be the big ball game anymore, whether we want to or not. . . .

Coming down here tonight I was thinking about the meeting over at Hardin last night, and some comments made there. They remind me of a time that I was living over at Ashland. There was a storm come in . . . and we were without power for seven days. Damn, that was a catastrophe in my house. Kids couldn't turn the TV on; they'd kick and stomp around . . . and they'd go in their bedroom and flip a switch on the radio; it wouldn't work. We didn't have water for our bathroom, so we'd have to go hunt up an outhouse someplace. Fortunately there were still a few around Ashland. We had a kerosene lamp there, and they bucked, kicked, and hollered about how sorry the light was from this kerosene lamp. Hell, I grew up with that. Some people call that the good old days, having to run out to the biffy and all this good stuff. I've got some real fond memories of those days, but boy I wouldn't go back to 'em for nothin'. I wouldn't if I could, that is. None of us can go back; we've got to work for tomorrow.

I was at the meeting in Bozeman, the one at Great Falls, and one at Helena. And there were quite a number of the younger people were making particular reference to not being able to see wildlife out their front door and all this sort of thing. Well, I can sympathize with them, but, ah, those days are gone. It's just like a speaker here awhile ago said something about. I can't quote him exactly, but it had something to do with numbers. And people.

[*Long pause.*] I lived on the reservation most of my life. Not all of it but most of it. And we Indians haven't had it very damn easy either. In fact, we've had it pretty tough, some tougher than others. But fortunately I can say that the Northern Cheyenne people, with particular reference to them because that's where I'm from, are starting to get their feet on the ground, and people are starting to listen to them. They're starting to do things. Even the Great White Father back in D.C. is starting to pay a little attention. Instead of telling us what we need. We're starting to get into a position now where we can tell him what we need and want. It's not 100 percent yet, but it's, it's coming.

So, with that, I don't want to take up any more of your time other than to say that I think phases Three and Four are needed. Thank you. [*Applause.*]

Bill Parker

[*In his haste to pack all he had to say into the time allowed, loosely, by Mike Moon, Parker stumbled over his opening lines, but after that it all poured out, home brew that had been capped too long.*]

My name is Bill Parker. I am a citizen of Rosebud County. I live on the Northern Cheyenne Reservation near Ashland, and I'm a member of the Northern Cheyenne tribe. It's a matter of historical principle that nothing is wrong—Nothing is wrong in pr—Is. It is a matter of historical experience that nothing that is wrong in principle can be right in practice. People are apt to delude themselves on that point, but the ultimate result will always prove the truth of the maxim: A violation of equal rights can never serve to maintain institutions which are founded upon equal rights. Secretary of the Interior Carl Schurz. 1878. . . .

That was the year the Northern Cheyennes left the prison camp in Oklahoma where they were sent and came back to Montana. . . .

[Today] we are faced with the prospect of trying to put ourselves together while suffering from lack of wisdom and leadership in the past thirty years. We know beyond a shadow of a doubt the Northern Cheyenne people, as with the gold in the Black Hills, as with the buffalo, so now together with the coal in the Powder River Basin are asked to offer ourselves up as the first sacrificial victims to this unwarranted invasion upon our land. . . .

I cannot speak here from official capacity or from a knowledge of the myriad legal issues, nor from the standpoint of the scientific pages of testimony which deal in quantities and qualities. However, a crime was committed even after Fort Robinson and the indictment hangs ever more heavy since before the turn of the century.

From what I am able to understand the Cheyenne people have a longevity to look forward to somewhere in the mid-forty years of age. The U.S. average is now in the seventies. That gives us a lifespan one half the national average. . . . We have just brought TB to a halt here. The respiratory ailments which beset Cheyenne people and others who have lived around us due to the dusty conditions we live in makes the epidemiological factors more virulent than any study has approached today. If we had had these conditions we live in with before and nothing was done about them—if they're still with us with the advent of heavy industrialization and nothing has been done to correct them, what can we look forward to in all reality?

A paradox exists in Billings, Montana. Just a few blocks from where this impact statement was put together. Surrounded by two oil refineries, two stockyards, two slaughterhouses, a sugar beet refinery, a sewage disposal plant, a coal-fired steam-generating plant, a soot-blackened and graffiti-covered cliff and a polluted river, and situated on what is politely known as a landfill but better known as a garbage dump, is a park. Dedicated to the Northern Cheyenne people. By the same people who for personal gain will want this travesty to expand as it will to actually surround our reservation at the very least. What can we expect from people who think so little of us? We have a miniature indicator of what is in store for the Northern Cheyenne tribe for all eyes to see. In Billings. Take a look as you drive by it, folks. No charge.

And these last remarks are my wife's side of the ten minutes, see. She's home pregnant and she can't get here tonight; she falls asleep and rolls over on her side and I have trouble getting her up.

I can't help but feel the effect on the human element is grossly understated from both standpoints of indigenous and influx population. I wonder if anyone has ever stopped to ask the workers in Colstrip . . . Wouldn't you rather have this plant built somewhere on the West Coast? Where you could relax on your days off and enjoy the things your money earns? Rather than have to cramp into a trailer camp, drink bad water, and have our kids slug it out in the overcrowded school surrounded by people who do not want to, but cannot help but resent you?

[*While Parker spoke there was noise in the hall. It bothered him. The proponents of the plant had shepherded their people out, he felt, to keep them away from the power of his words. Their voices clattered into the room.*]

If this electricity were to be used in heating homes, hospitals in ghettos, then one could at least understand the tradeoff in human life. But when one realizes that most of this will be used in oil refining, aluminum and copper processing, or wasted, then there is no justification. Send them the coal if they need it so badly, and save all our already badly used lives. . . .

Thank you. [*Applause.*]

Vic Jungers

[*"I went to Forsyth and I sat there all evening and listened to everybody and I never put my name down," Jungers remembered. "Then about midway through I felt that I had to commit myself one way or another. I think it was the priest and the McRaes that irritated me the most. And then about ten o'clock I asked [Moon], I said 'How many's on that list?' There was a long list. And I said, 'Well, I have to speak. I just can't sit here any longer.' So I got a card from him and put it on the bottom."*]

My name is Vic Jungers, from Colstrip. I am employed by Bechtel. I'm an ironworker. Can you hear me back there? I usually scream most of the day, so I hope you can hear me—May I use the mike?

I've attended three of these meetings. And as I go to more and more meetings I find myself more and more involved. Not with cold logic but more with emotion. . . .

We've heard a lot of talk about taxes at the last three meetings . . . I say that, you know, we can change the law. You people, and myself, have lived here for a good many years under the present taxation system. Now it is not satisfactory. Maybe you and myself should change our representation. But in this part of the state historically our representation has come from an agrarian society and as I perceive that society as expressed here tonight that means to preserve exactly that: the agrarian way of life. We have heard the theory, the greatest good for the greatest number; that's been equated with socialism. And this particular gentleman went on to say this is what happened to the Indians. I say to you, if it didn't happen to the Indians, where would you be today?

Think about that.

We heard a little thing on Patrick Henry: Give me liberty or give me death. Today it's supposed to mean give me, give me, give me more. If that implication was that the majority of people want Three and Four don't give it to them because I want liberty,

liberty in this instance being equated with a certain way of life, an agrarian way of life, this is the same selfish self-interest you people accuse the Montana Power and large conglomerates of.

[*A smile.*] Ah, maybe I'm a socialist, but if this is socialism, I say, let's have a little more. I think I can decide when I've had enough. . . .

Speaking of self-interests. We all have self-interests, even the English teacher who spoke from St. Labre had the self-interest. She very poetically described the sweet and tranquil pictures of this prairie and the Montana ecosystem as we now know it. . . . To paraphrase her, "I want to give the future the peace and security that I have known as a resident and teacher in Rosebud County." Ah, not even thinking about how or why she happens to be teaching in St. Labre. Ah, her very dream came about by the destruction of another people's dreams. . . .

I, too, would like to address myself to the clergyman. I would like to remind you that historically the church was and is oppressive, and repressive, supporting their own brand of status quo. You, too, have to destroy dreams of another culture only to replace them with dreams of your own.

I'd like you to think about that.

Ah, I'll get down to the more emotional thing, where I get more and more involved and find myself taking a more intransigent position. . . . I see this dichotomy. . . . It's basically boiled down to the rural versus urban, industry versus agrarian. And I happen to favor the industry a little more than I favor the agrarian although I realize that we have to make some kind of viable compromise between the two. But this compromise will never come about unless we have some change. . . . Change to me means building Three and Four. . . . [*Applause.*]

Don Bailey

I've nearly forgotten what I was going to say, I've been here so long. I'm Don Bailey, I live south of Colstrip, I think. I'm not sure my wife would agree with that—I have been on this tour now for some time, and it has affected me in a different way, too. I think that basically each and every one of us here tonight and all those that have left are identical in one instance. We're self-preservationists. And this is only human nature, and I think we can all accept that. I think that the arguments for and against coal development, or the construction of Three and Four, depend on one's own personal philosophy. . . . And I think that for every argument for there is an equal one against. And vice versa.

[*Outside there was a long, mechanical wail, and the night's last coal train rumbled east. And in the room the ring of the crossing bells and the rattle of the steel on the rails went on and on until at last it seemed an echo of itself, endlessly repeated. Bailey talked over it; the noise became a part of his speech.*]

I have seen something develop as I've been on this—circuit, as we might call it. It disturbs me. I've seen a polarization of the rural community toward the working community and vice versa, and I don't think that this is going to lead to a healthy situation. . . . We've heard attacks against principle, philosophy, and person here tonight; I've heard it before, at other meetings. And I'm beginning to question the value of these hearings, as they are conducted. . . . I think it's important to have public input, but I wonder if it wouldn't be better if we all took the time that we took to come here and really dug into the issue, wrote a statement, sent it in to the department for them to analyze at their discretion and come up with their own decisions. And to eliminate this animosity and this nitpicking that's been going on, to alleviate this thing that I see building here. . . . Thank you. [*Applause.*]

41

Leaving the courtroom, walking down the wide stairs from the third floor, Sheila McRae was stopped by a woman she had never seen. "I think she was in her twenties," was all Sheila remembered later. The woman stopped her and shouted in her face.

"You! How can you say things like that?" Her eyes were malignant, tearing, like claws. "You call us drug addicts—"

"I didn't say that," Sheila began. "I—"

"What right do you have to tell us how to live? What makes you so important?"

In the turmoil on the stairs someone grabbed the woman's arm and pulled her away. Sheila stood there, eyes glistening, watching the space where the woman had been. Then she walked swiftly down the stairs, where friends met her and talked the fear away.

Tom and Donna Wimer, Charlie and Barbara Wallace went to Buff's Tavern, the only place in Forsyth open for dinner after midnight. The place was crowded; the light was bad. In the crush of people working their way into the dining room, a large man stopped Barbara.

"Hey," he said. "You're the one who got up there and told all those lies." She tried to get around him. He stood in her way and looked over her head at her father and at Charlie.

"Sure," he went on. "You're the one that thinks you're going to tell on the big, bad power company."

She got around him and started to work her way to the door. There was a sound of male laughter behind her. She could see candles on the tables in the dining room, but this place was formless, dark.

"Hey," the voice said behind her. "I'll tell you, little lady, you aren't going to have to worry about working for them bad guys anymore."

176

The sound of his voice disappeared in the laughter. The Wimers and Wallaces went in to the dining room and sat down.

Vic Jungers drove back to Colstrip with his girl friend, Connie, who got scared of the ice and nearly spun out near town. The Olmsteads drove back with several other executives, and dropped one of them off just outside of town where several cars returning from the hearing had run off the road. Wally McRae gave Nick Golder a ride down to Colstrip, but let him out early to help a frightened Colstrip couple who had also gone off the road.

Martin White went home, and in due course caused a letter to be written to the St. Labre Indian Mission. So it happened as Brother Ted Cramer had feared: The mission never again took an active role in the controversy over the power plant at Colstrip, which swiftly reached out to engulf the Northern Cheyenne tribe.

Mike Moon packed up the tape recorder and went out into Forsyth. To his surprise he found few of the people who had come to the hearing in the bars. But outside the Joseph Hotel he was propositioned by a whore. He turned her down.

And nobody listened to the tapes of the meeting in Forsyth until I came along almost two years later and found them in a closet in Helena. But by then the first two stages of the power plant were running.

Part 3

Part 3

42

The largest strip mining shovel to be erected in Rosebud County took its first steps on Wednesday as Western Energy applied power to its latest piece of equipment.
—Forsyth Independent
December 19, 1974

A coal industry-sponsored poll funded by the Montana Coal Council, Montana Power Company and Franklin Real Estate Company said 59 percent of Montana residents favor more industry. . . .

A revised public opinion survey conducted by the Montana Department of Natural Resources, shows 53.3 percent of Montana residents flatly oppose construction of coal-burning generation plants. . . .
—Bozeman Daily Chronicle
January 9, 1975

Montana's five newly installed public service commissioners have gone on record in opposition to proposed power generation plants at Colstrip.
—Livingston Enterprise
January 10, 1975

The Montana Department of Natural Resources Tuesday recommended to the state Board of Natural Resources that the application of five utilities to build Colstrip Units Three and Four be denied. [The reasons include: There is no need for a 1,400-megawatt plant in Montana within the decade; that there are reasonable alternatives to the proposal; that the plants do not represent the minimum adverse environmental impacts for the state of Montana; and that "the department is not persuaded that the obligation of Montana to the rest of the nation or the Pacific Northwest necessarily goes beyond making energy available through reasonable alternatives." If plants like Three and Four, with their emphasis on exporting

181

power, are constructed in the state, the department said, "Montana could become a boiler room for the nation."]

—*Bozeman Daily Chronicle*
January 29, 1975

Officials of several utility companies warned Tuesday that loss of proposed electrical generating plants could contribute to a long range power shortage.

—*Miles City Star*
January 29, 1975

The public hearing [called by the Board of Natural Resources to receive testimony on Colstrip Three and Four] will begin March 10 in Bozeman.

—*Miles City Star*
February 7, 1975

Convinced that the final hearing on Colstrip units Three and Four . . . will take six to eight weeks, the Montana Board of Natural Resources voted late Monday night to have someone else hear it for them. . . . [The board also] voted to postpone the hearing until April 21.

—*Bozeman Daily Chronicle*
March 11, 1975

Power companies picked up the tab and most of the time to present their case on development during Monday's guided tour of Colstrip conducted for more than 200 persons attending the Western Governors' Conference. . . . "I think it was blatantly unfair to have an energy tour during the conference when the hearings on power plants Three and Four are going to be conducted soon," [said rancher Wally McRae]. "On my bus I at first almost had to be rude to get an opportunity at the microphone. . . ."

—*Billings Gazette*
April 1, 1975

Saying critical public decisions should not be made "By computers programmed by unidentified public servants," a Montana judge ordered on Tuesday that the state Board of Health conduct public hearings on health matters critical to the future of a huge power complex.

District Court Judge Gordon R. Bennett rejected the contention of the state Department of Health that the department should be the final governmental agency to decide whether the proposed Colstrip Three and Four complex would meet legal requirements for air and water pollution.

—Bozeman Daily Chronicle
April 30, 1975

Carl Davis said the State Board of Natural Resources would begin hearings on May 20 on applications for Colstrip Units Three and Four.

—Butte Standard
May 6, 1975

Montana Power Company's president, George O'Connor, was on the stand for seven hours Wednesday. . . . Under probing he insisted that Montana Power Company has no plans for construction of generating plants beyond Three and Four.

—Havre Daily News
May 22, 1975

. . . O'Connor said he had no idea why a MPC official in January, 1973 had not divulged plans for . . . Colstrip Units Three and Four.

—Miles City Star
May 23, 1975

The Colstrip generating project was designed indirectly to sell Colstrip coal to Montana Power's Pacific Northwest partners in the form of energy, Leo Graybill [attorney for the Northern Plains Resource Council] suggested at the Board of Natural Resources hearing.

O'Connor replied: "We sold it to them at Colstrip when we couldn't sell it to them at Portland."

—Bozeman Daily Chronicle
May 23, 1975

State hearings to determine whether two coal-fired generators should be built will last until 1977 unless the present pace increases, hearings examiner Carl Davis said. "I told them [the attorneys] we ought to develop a better emission control system in this room . . ." he said. As of Tuesday, the 24th day of the hearings . . . the attorneys had completed questioning only five of the scheduled 91 witnesses.

—Butte Standard
June 26, 1975

[*A few days later Davis suspends Board of Natural Resources' hearings to allow the Board of Health to hold its own hearings.*]

A federal energy administration team will look into the Colstrip generating proposal and attempt to determine what can be done to step up the process.

—*Butte Standard*
July 5, 1975

. . . Rosebud County Sheriff Jim Schiffer, speaking to a forum at Dawson Community College . . . said that felony arrests have jumped from 18 in 1971 to 319 in 1974, and that there has been more than a 42 percent increase in civil action procedures in the past 5 years . . . [He also said that] four of his officers have been attacked and injured in bar fights [since coal development began] and that Rosebud County now leads the state in per-capita crime.

—*Terry Tribune*
July 31, 1975

Colstrip Unit number One is expected to turn for the first time later this month.

—*Forsyth Independent*
August 7, 1975

Steve Brown, counsel for the state Department of Health, said Thursday that "Montana does not need a $1 million white elephant." On the basis of upcoming testimony, Brown said that the health department "must recommend that certification not be granted on the issue of compliance with air quality laws and standards. . . ." Referring to a now-recessed Department of Natural Resources phase of the hearing, Brown said witnesses testifying on behalf of that agency "established serious deficiencies in the pollution-control system proposed for Units Three and Four. "Simply put, they won't work," he added.

—*Miles City Star*
August 7, 1975

Indications are it will be at least mid-October before the state Board of Health can hear closing arguments on the controversy over Three and Four. . . .

—*Miles City Star*
August 29, 1975

The Board of Health's 53-day hearing ended . . . after . . . a meeting of less than an hour Tuesday.

—*Miles City Star*
September 17, 1975

The State Board of Health tentatively agreed last week to give preliminary go ahead to proposed Colstrip power plants Three and Four after they heard final arguments two weeks ago. "Some, at the outset, felt denial might be necessary, but they changed their mind," said board member William Spoja, Jr., Lewistown.

By the end, Spoja said, "I don't believe anybody believed denial would be the action to take."

—*Butte Standard*
November 23, 1975

Preliminary test results released Friday by the state health department show Montana Power Company's coal-fired generating plant in Billings is in violation of state pollution control regulations.

. . . Earlier this week the bureau refused to disclose the test findings. Steve Brown, department lawyer, said on Wednesday that release of the information at that time could bias the health board's decision on the Colstrip plants. Tom Scheuneman, a Chicago lawyer representing the Northern Cheyenne Indian Tribe, said the . . . tests demonstrate "the Montana Power Company does not know how to operate within the law."

—*Miles City Star*
November 23, 1975

Portland General Electric Company has filed a report with the Board of Natural Resources that appears to support the Department of Natural Resources' contention that it is cheaper to ship coal . . . than to burn it in Montana and transmit electricity to the Pacific Northwest. . . . The report, filed as part of [the Board's recently resumed hearings on Colstrip Units Three and Four] says it is cheaper to haul coal by rail to a plant to be built at Boardman, Oregon, than to transmit electricity generated at Colstrip.

—*Miles City Star*
January 26, 1976

Timothy DeVitt, a vice-president of a Cincinnati consulting firm, said material submitted by the five utilities show the system [for pollution control for Units Three and Four] with a guaranteed average sulphur dioxide control efficiency of about 40 percent. "Given the present state of the art, it is possible to design a system to remove at least 90 percent of the sulphur dioxide emissions entering the air . . ." he said.

—*Daily Inter Lake* (Kalispell)
February 20, 1976

A spokesman for the Montana Power Company indicates the firm is thinking about building a 350-megawatt coal-fired power plant beyond the current ones [at Colstrip]. The first indication that MPC is looking beyond Colstrip units One, Two, Three and Four came in a 10 year development plan filed April 1. . . . The plan also says the utility company is contemplating participation in one or more coal gasification plants.

—*Bozeman Daily Chronicle*
April 19, 1976

. . . The Montana Board of Natural Resources will hear closing arguments today and Thursday on whether to allow construction of two coal-fired power plants at Colstrip. . . .

—*Spokane Spokesman Review*
May 19, 1976

DECISION ON COLSTRIP PLANTS SEEN POSSIBLE ON JUNE 25
—*Spokane Spokesman Review*
June 11, 1976

Members of the state Board of Natural Resources gathered in secret in a Helena motel room Thursday for an unannounced meeting to discuss their upcoming decision on whether to approve construction of two proposed power plants.

The meeting was not publicized in advance, and board members seemed surprised when a reporter dropped in. . . .

[Chairman Joseph Sabol, Bozeman] contended that the gathering in his motel room was not a meeting because "all the board wasn't there and we weren't . . . voting or taking a vote." Cecil Weeding, Jordan, had not arrived yet, but the other six members and hearing examiner Carl Davis, Dillon, were present.

Another board member—Wilson Clark, Billings—added "The open meetings law has a lot of real merit, but what it doesn't recognize is that a public body that has to make a decision has to thrash it out."

—*Butte Standard*
June 25, 1976

. . . The Board of Natural Resources gave conditional approval to the $1 billion project . . . Friday. [The vote was four to three.] Gary Wicks [director of the Department of Natural Resources] will not initiate an appeal. "We don't have the

money," [he said]. "The decision has been made, and it's time to live with it, I guess."

—*Butte Standard*
June 26, 1976

... On Friday Allen Rowland, chairman of the Northern Cheyenne tribal council, announced the tribe's petition for a redesignation of reservation air quality from Class II to Class I [under the Environmental Protection Agency's regulations for prevention of significant deterioration of air quality. Class I is the highest quality air].

—*Forsyth Independent*
July 8, 1976

The Northern Plains Resource Council and the Northern Cheyenne Indian Tribe have appealed the state's decision to allow the construction of two additional coal-fired generating plants at Colstrip. Appeal was filed Friday in district court in Helena. The appeal charges the boards of health and natural resources made substantial errors in granting conditional certification.

—*Daily Inter Lake* (Kalispell)
August 24, 1976

... The state of Montana and Environmental Protection [Agency] officials discovered that while Montana Power Company attorneys were telling the state that Colstrip Three and Four would comply with federal [prevention of significant air deterioration] rules, the utility was trying to persuade EPA that the Colstrip expansion would be exempt from PSD (Prevention of Significant Deterioration).

—*Bozeman Daily Chronicle*
August 24, 1976

... According to a recent report in the Missoulian *newspaper, Montana Power Company attorneys told both the state Board of Health ... and the Board of Natural Resources that Units Three and Four would comply with federal clean air rules regarding prevention of significant deterioration. But, at the same time, MPC attorneys were trying to convince the Federal Environmental Protection Agency that Units Three and Four should be exempt from PSD standards because construction on units began before June 1, 1975.*

—*Butte Standard*
August 27, 1976

The first plants at Colstrip are to be dedicated tomorrow. Three thousand to 5,000 people are expected to attend, including the governors of Montana and Washington.

—*Havre Daily News*
September 17, 1976

More than 3,000 braved wet, windy, and chilly weather Saturday as an assortment of dignitaries dedicated the $294 million Colstrip generating Units One and Two. The ceremony was held in the windswept parking lot of a new shopping center that grew with the two coal-fired generating units.

—*Miles City Star*
September 20, 1976

On Saturday the titans of two states pointed with pride to the monument they had built to the future of America on the prairies of Southeastern Montana. . . .

—*Forsyth Independent*
September 23, 1976

43

I awake late in the morning to a huge and violent light flashing on and off in my trailer. It comes and goes totally without warning, without rhythm. It pounds on my closed eyes. Booom, dark, Boom-dark-boom, Boooom. Dark. I get up and look out the window. It is a beautiful, sunny day, but cold. The wind is blowing hard from the south, an unusual direction. The two columns of cloud from the stacks of the power plant race directly across Burtco Court, and my trailer is at the edge of the shadow. As the wind-blown cloud races overhead the sun ducks in and out behind it and brings this turbulent light to my home.

I have been living in Colstrip now for a week. I moved down here from my home in Idaho to write this book, and on a side trip to Helena I picked up the names of those eighteen speakers at the Forsyth hearing at the end of 1974, and made a copy of the tape recording. Mike Moon, who was still working for the Department of Natural Resources, guessed that no one had ever listened to the tape. He certainly hadn't; and when I found the tape with a group of others in a box deep in a closet, it was marked "Hardin." Some of the other tapes were missing. "I don't think public comment at the hearings had any influence with the board, really," he said wryly. "Very, very minor influence." He didn't seem to mind the futility with which that comment invested his efforts of two years before. He was, after all, in government work. So in the past several days I have been sitting here listening to the tape, transcribing the forgotten testimony, hearing the mutter of the crowd in the background and the emotion in voices of people I have not met yet. Surely this meeting mattered somehow, somewhere—if to no one else, at least to the people who spoke?

I make some morning tea and go over my notes, my back to the flashing sunward windows. I have been reading in news clippings about some of the changes that have happened here since December 18, 1974. The most important of them is not really a change at all, but is rather a startling consistency that no one would have predicted two years ago. It has now been over twelve hundred days since Montana Power Company and four other utilities applied to the state of Montana for permission to build Colstrip Units Three and Four, and construction has still not begun—at least practically. The companies are in the process of arguing in court that it began before June 1, 1975. Other impediments to construction are the Northern Plains Resource Council and Northern Cheyenne suit against the state for holding what they call improper hearings, and the Northern Cheyenne request for the Environmental Protection Agency to designate their reservation a sanctuary of Class I air, officially the purest air in the nation. No one knows how that will affect the power plant if it is approved. So the space just east of Units One and Two remains empty, waiting in a very expensive limbo. The power plant has not yet won, or lost, its war.

I sit down at the little plywood desk I have built into an alcove in the trailer and look out the window, a window I have tried to insulate with a sheet of plastic taped over the inside of the glass. What will I learn tonight, when I begin, with Wally McRae? Again I feel less of a reporter, more of a character, caught up in something I can't control. Whatever it is I am a part of here, it is not yet over. I feel I am trying to describe a ship—its length, tonnage, speed, and magnetic course, from somewhere inside—the boiler room?

My view, distorted by ripples in the plastic, is of twenty-five other trailers marching up the slope. At the top of the slope is a little ridge of old spoil, and a barbed-wire fence, then prairie. Over this, too, lies the racing shadow.

44

One hundred and sixty-seven sections [107,480 acres] of land in Southern Rosebud County have been targeted for coal development by companies responding to a call for nominations by the U.S. Bureau of Land Management. However, the BLM is not saying what company wants to develop what land.

—*Forsyth Independent*
October 21, 1976

The Montana night is complete, deep as its past. The sun has disappeared behind the invisible Rockies, and the red hills have given up their glow to a cold sky. The hills are black now, and the sky is a deep blue-tinted gray in the west, shading darker around to the east where the hills and the cloudy sky mingle in a uniform darkness. As I drive down the Rosebud road on my way to Wally McRae's house, I am engulfed in night. Only the short tunnel of my headlights and the very occasional yard lights of ranch homes along the road—Jim Snider's place, Duke McRae—remind me that this is the twentieth century.

McRae's home, an old frame farmhouse, is on a little knoll above the Rosebud. I get out of the car and shut the door. There are soft thumpings and breathing from a corral across the yard. Horses. One yard light reflects off cold, dry snow. At first, except for lights in the windows, this is the only light I see, clear out to the western hills. I stand in the yard for a moment, absorbing this wide isolation. Will this become a part of an historical memory; will I be telling younger people in fifty years that I once visited a home where you could see no other lights from its windows?

But the purity of my illusion is destroyed. Off to the north there is a flicker on the horizon. Over beyond a ridge five miles away, the symmetry of this bowl of night is broken by a rectangular puff of light.

191

Flash, flash, flash. Every five seconds the horizon, whose line between sky and land is smeared to obscurity by the darkness, is lit and silhouetted by a short gleam. This light ticks, bright enough to see but too quickly gone to grasp except as an echo on the retina. Flash, flash, flash. It becomes established in my vision; as I turn toward the house it remains ticking in the edge of my mind, like the flutter of astigmatism in the corner of my eye. It is the strobe lights on the stacks of the power plant at Colstrip.

McRae welcomes me at the door. I hang my coat in a vestibule jammed with coats, boots, and cowboy hats. He invites me to dinner. Roast beef, potatoes, gravy, fresh milk.

Wally McRae has the small eyes of a shark and the scars of a middle-echelon predator accustomed to fighting for his food. After dinner he shows me a photograph of himself that he keeps stuffed in with the speeches, fact sheets, environmental reports, and bits of poems that are stacked in a pile beside his chair. It is an eight-by-ten, mounted on cardboard, a head shot from the side, backlit so the face is in shadow. There's the dark hair, the sideburns, the heavy brow that looks as·if his brains pushed his head out from under his hat, the bristling moustache, the creased jaw, and the thick nose, an inheritance from his father. In the photo the brows are low, and by a trick of angle the eye has no pupil and stands out, white and empty, in a face heavy with deliberation. McRae lets me look at it for a moment, while he gazes off in another direction. Then he says: "It makes me look as if I'm strangling somebody—slowly." He sits at the end of the kitchen table, and appears to be looking past the kitchen to some far horizon, his face somber. Then his narrowed eyes flick sideways at me and then back to deadpan, and the effect is of an oil portrait winking. "I like it," he says.

Ruth, McRae's wife, brings coffee. McRae stubs his Lucky Strike out in an ashtray shaped like a spur. There is a faint air of ritual in the room. The coffee, the packing off of the three children to homework and bed, Ruth's quiet attention to dishes. It is the beginning of an interview. McRae settles back in the chair, sighs audibly, watches me untangle the cords of my tape recorder and turn it on, and waits for the familiar questions that I will produce as inevitably as if I were a machine myself.

I'm hardly the first reporter to come to his house on the Rosebud. I'm so late I'm almost a camp follower. The main army has already been through. Douglas E. Kneeland of the *New York Times* came and sat in this room, remarked upon the longhorn cowhide draped on the couch, and called McRae a "sun and wind-reddened rancher." In a story headlined TO STRIP RANCHES FOR COAL: DIFFICULT CHOICE, Kneeland described McRae staring into a cup of thick black coffee, which is what he is doing now. "I don't think the people coming to Colstrip understand why I resist coal development," McRae told Kneeland. "They think I'm a miserable son of a buck. They can't see why I resist selling this ranch for more than the land is worth. They think I'm just an obstructionist. . . . My blood just runs in a different direction than theirs."

James Conaway found his way down here, too, doing a sad, contemplative article for *Atlantic Monthly* called "The Last of the West: Hell, Strip It!" "McRae," Conaway observed, "is known somewhat affectionately as a 'super radical,' a fact that saddens him." "Coal mining is an inexorable force," McRae told Conaway, no doubt while staring at a cup of thick black coffee. "Now an acre of coal is worth between $50,000 and $500,000. There aren't enough people in the state of Montana to stop the mining."

Now it's my turn. McRae picks up the mug and sips from it. He looks at me out over the rim, and I unpack my questions, set them up on the table like paper ducks, and he knocks them down one by one. Age, family, background, number of acres, number of cattle (six hundred), focus of his anger (Colstrip Three and Four), reasons why he keeps fighting. And as the preliminaries fall in sequence McRae's mood seems to change from one of almost flippant attention to details into sadness, as if he requires the preliminary questions to knead out his growing skepticism at his own facile performance, and release him to say what he really believes without being irritated by the echo of his own familiar words. He is a night person; alone, he pounds out letters or speeches or poetry way into the small hours, his thoughts cleansed and intensified by the silence. Now, as we talk late, he reaches for his power.

"If you go back to the days of the Homestead Act," he says, starting slowly, quietly. "There were two main pushes behind it. The federal government and industry. The industry was in the railroads, because they wanted to haul people out and products back. And the federal government wanted it to develop and settle the West. And they brought these people out here and dumped them off on a marginal piece of land that wasn't big enough, and it was a terrible tragedy that was perpetrated on a bunch of naive people. The same thing was done in the case of the Indians. The federal government and some mining and ranching interests decided that they wanted to settle and populate and broaden and expand the tax base. So they came out here and did the same thing to the Indians."

He pauses, sips at the coffee. Now his voice is deeper, wider, he's the cowboy orator, standing slender and straight at the rostrum, reaching back for the far corners of the gym.

"And now we've got the third thing. And the same people are doing it: The federal government says that there is an energy crisis, we've got to become self-sufficient in energy, and some large energy conglomerates are saying we're going to broaden and expand your tax base, we're going to settle the West, we're going to disperse population, and it's going to be an unmitigated good thing. It's just another travesty, perpetrated by the same damn people that have been doing it forever."

Silence. McRae pulls a kitchen match from his shirt pocket, strikes it with his thumbnail, and lights up a Lucky Strike. He seems deflated. Maybe the echo returned. Perhaps he hears the words repeated on my tape recorder, then repeated in my writing, stepping down into the distance like the images in facing mirrors, getting paler all the time. He sighs.

"You know," he says, "one of the things I do a lot is get into the question of am I doing what I am doing, which I think is a good thing, for the right reasons. Am I hassling with Montana Power Company because I think Montana Power Company is doing something wrong that ultimately really will hurt me, or am I hassling with Montana Power Company because they're the big guy? Am I fighting the identifiable power structure because of frustration on my own? Maybe I'm satiating that frustrated dramatic bent. I don't know.

"There are a lot of old jokes about the code of the West. And it is true. There is a defnite sense of propriety, and it is hard to explain, but there are things that you do, and things you don't do. And that's part of what we're fighting for. We're all convinced that the thing we're doing is the right thing to do. But if we're going to use the same tools that the energy companies use, then we've lost the battle. Because the rules they operate under aren't compatible with openness and frankness and honesty, like we're used to doing. People will come from an energy company and they'll tell you something, and you'll believe it.

And, hell, it's not the truth; it's merely expediency. And yet if we understand the situation and if we deal that way too, then most of the things that we are fighting for, which is a way of life, are already lost."

There is weariness in the voice, a weight of pessimism. Caught by that mood, I ask the natural question: Do you really think, given all the power, the money, the historical momentum that stands behind coal development, behind the power plant, that there is much hope of stopping it where it stands?

He studies his cigarette and stubs it out. A last small puff rises from the ashtray and dissolves gently in the air.

"Yeah," he says eventually. "I think there's a chance of stopping it." His voice is still quiet, but it is quiet now because he seems to be restraining it, not because he's tired. His narrow eyes flick over to me again.

"Yeah," he says. "I'm going to give her a whale of a fling."

It is 1:00 A.M. It is ten degrees outside as I leave the house. I say good night, breathing frost, and walk slowly across the yard to my car. The thin snow is dry and light and shallow; it makes a small creaking noise under my feet. It glitters in the reflection of the one yard light. The clouds have lowered, and to the north the strobe lights on the stacks at Colstrip beat against them, *flash, flash, flash,* and against the ancient Montana night.

45

"The grounds for our determination . . . [that Colstrip Three and Four are covered by new Environmental Protection Agency clean air rules] were that there had been no on-site construction, that no authority to commence construction had been obtained from the state, and there was insufficient evidence to find that the units should nonetheless be considered as having commenced construction because the company could otherwise sustain substantial unavoidable losses if they had to cancel or modify contracts entered into prior to June 1, 1975."

—John A. Green, EPA
Regional Director, quoted in
Forsyth Independent
October 28, 1976

Mike Hayworth lives in one of the new houses built along curving streets to the north of the old Colstrip. From his street there is a good view this month of a dragline marching slowly along a ridge to the west. This dragline, the same vast machine that took its first steps—literally—the day after the Forsyth meeting, flings a bucket out at the end of a long boom and scrapes earth and rocks from the surface of the coal, then heaps this overburden to one side of the trench it thus creates. Its size makes it seem to work with a stately deliberation, swinging, dumping, casting its bucket, dragging, and swinging again. It works twenty-four hours a day; from the town you can see the lights at night, tracking slowly back and forth through the darkness like the lights of a ship anchored on the tide.

I find Mike Hayworth putting on tennis shoes in his living room. The house is large and clean. There is a lawn outside. Hayworth is stocky; he wears thick wire-rimmed glasses. When he starts to talk he has no reticence at all. When I ask him how he developed his enthusiasm for the power plant, he tells me about giving tours of the mines back in the early 1970s, while he was still teaching.

"My initial pursuit into it was for a part-time job," he says, "and from that time I was looking toward . . . Gee, this is a great career possibility. One summer I was tour guide and at this particular point in time I studied and got to know quite a lot about the mining issue and reclamation and so forth. And I think that at that point in time I was clearly, you know, on the side that said: 'Hey, let's not say that this won't work because here I am standing on grass up to my waist and I know a lot of places in Montana where it just doesn't grow like that or you just don't get the tons and tons of vegetation growing in spots like this.' "

He talks rapidly, as he did at the meeting in Forsyth. He tells me about his letter to the editor of the *Billings Gazette*, the letter that brought him out into the public eye and made him want to keep participating. I get the feeling that at some point there Hayworth was beginning to enjoy the role of spokesman. After his response was printed, about the same time as the Forsyth hearing, there were replies to it, and Hayworth was all ready to accelerate the dialogue with another letter. But something happened.

"I would say that I really wanted to go ahead and continue and answer what was said," he says, and the pace of his words slows. "But I would have to say my wife influenced that quite a bit too because I may have went ahead and done it if it had just been . . . but she influenced me quite a bit because it isn't just me living here associating with people. My wife and family are living here and associating with the people, too."

After his letter and his speech, Hayworth may have thought he was beginning a new step in the direction of a public life. He was about ready to be recognized as an activist. One more hearing would have done it. "I hope you don't quote me as being an activist," he says, a little hopefully. "But I did take an active part in this."

But after Forsyth he didn't speak again.

"I think the overall thing was I said things that brought items to light that were actually not in the correct context," he says. "And that gained me some enemies. No matter how true or correct things were that I said, if people didn't like what I said, and don't like me because of that, that stays around a long time, and I don't know, I have—I would say as a result of this, I have not been so vocal on some things. I see these people at ball games, and, well, that's coming back to haunt me again. You know, the hearings ended at ten, eleven, twelve o'clock that morning, but I continue to live on at Colstrip."

There is a momentary silence. Hayworth is getting ready to go out. So I ask him one more question. I have heard that his brother, Pat, who had testified against Three and Four, had quit the cowboy life and had gone to work for the mine. Why did he do that? I ask.

Mike Hayworth brightens, becomes almost cheerful. Here is his one victory of that unfortunate night.

"You'd better ask him that," he says, and laughs. "Yes, I think that question had best be answered by him."

46

In a hearing held last month in the federal district court of Judge James Battin in Billings, witnesses for [several] power companies testified that if construction does not begin soon, power from Colstrip Three and Four will not be available in time to prevent power shortages.

—Forsyth Independent
December 2, 1976

Pat Hayworth grins. "Well," he says, "it would appear that I—what would you call it—compromised myself. Is that it?"

I don't quite know what to say—agree or disagree. Pat Hayworth lives in a small trailer near a ranch house down on the Rosebud. He is a slender man, not large, with short, dark hair, a snub nose, and a shy, friendly manner. We sit in the small living room. He knew my first question was going to be why, after opposing it so vigorously, he went to work at the mine. So the answer came quickly. Now he elaborates.

"It was a combination of things. Mostly it's your depleted cow prices. You know, cattle's slow and just about any rancher can't afford to pay too many good wages." He paused. "Oh, and if it isn't me workin' over there it's somebody else. I might just as well get part of the fat of it, too. You know, I don't plan to do this for the rest of my life, minin'. "

I ask how much he makes.

"Ohhhh, well . . ."

How about by the hour, anyway?

"Nine bucks an hour."

What do you do?

"I run a coal shovel." Ah, I think, another heavy equipment operator. It must be a big change from rounding up cattle.

199

We divert the conversation from his work at the mine. He still wants to be a cowboy; his real goal is to manage a ranch for someone. And he hasn't changed in his opposition to strip mining and the power plant.

"To me it's a little scary sometimes," he says. "You know, they're taking that earth and disrupting it, moving it. They say that air is clean coming out of them smokestacks. Sometimes it *looks* clean, you know. But there's supposed to be all this sulfuric acid—or it turns into sulfuric acid when it mixes with water. Well, you might not know this for another five or ten years. . . . Some days it don't bother me. You know, I mean it isn't that big a thing. I know when I drive to work every morning or every day, you know I notice it, and I think about it a little bit, I b'lieve."

Pat's wife, Jo Ann, joins us in the living room. She is a teacher's aide at the Colstrip school. She listens quietly while Pat talks. For a man who is notorious as a quiet individual even among the relatively taciturn men of the Rosebud, he has plenty to say.

"We lived in Colstrip for a while, when I went to work down there. That's what struck me more than anything—when they threw, oh, heck, it must have been five thousand people. When they threw five thousand people out in the middle of the prairie and no place to shit. I didn't care about living in that close proximity with everybody else but, you know, the people were nice. I mean, if their garbage can tipped over and rolled over on your part of it, you know, they were nice about it. They'd come and pick it up."

Pat moved to Colstrip last January and escaped back out to this rural trailer in May. When he has the time he helps the rancher he rents from with his chores. And he waits for that ranch managing job to turn up.

"What this mining job has done for me is give me time where I can look around and wait. It's been a pretty good job for me. I mean my employer's been pretty good and I haven't got no complaints about the job, just as far as the work I do. I mean I won't say that I b'lieve in your strip mining or these plants; I still disagree with them puttin' in these generating plants, just as much now as I did then."

I ask him if this belief has ever resulted in conflicts with his fellow workers.

"Oh, no," he laughs. "The thing that surprised me most is when I went to work over there. The guys I worked with and around them plants ... they weren't involved in the debate. Seems like it was over most of them's head. Some of the guys wanted them to build Three and Four, a lot of them were against it, even that work over there. I don't know, it really surprised me."

Before I leave an interview I like to tidy up the details that I tend to forget in the beginning. Age (thirty-two), spelling of wife's name, spelling of children's names (one, Emmy.) Now I ask Pat just what kind of coal shovel he drives. How many tons fit in its bucket, how many cubic yards of coal?

Pat grins; there is a little return of shyness. He glances over at his wife.

"Well ..." he says, beginning to laugh. "It's got a number two written on the handle. It's a scoop shovel. You gotta keep those conveyors clean. I've told a lot of people that. They ask me what I do, I say I operate a coal shovel. Don't have to tell 'em it's five feet long and got a wooden handle."

Now I've embarrassed him, and I'm embarrassed too, though it doesn't often happen that someone else's revelation bothers me. But I can see him too clearly in the yard on the night shift, a tiny figure sweeping, shoveling away in the many-layered night shadow of the huge machines. The tin shovel clanks on the cement; the coal rattles in the bin. A cowboy's face and hands, snub nose, happy eyes, powdered black with the unburned soot of the mine.

47

The dragline is the arm of the power plant's local imperialism, reaching out into the land to consolidate territory for the plant's consumption. It is the battleship of the plant's fleet, six million pounds of steel armed with a ferocious long claw. Its boom is 325 feet long; in one scoop it can lift seventy tons. It walks: All that bulk rests on two fifty-eight-foot-long steel shoes that heave along, one at a time, taking seven-foot-long steps at just over one-tenth of a mile per hour. I have been aboard a walking dragline; it waddles backward, groaning and grunting as it moves. Behind it emerges slick, flat earth, steaming from the pressure and the friction of the passage of the huge machine. Ahead of it the earth is split asunder, and in the pit it makes powered shovels work feverishly to remove the coal, like hyenas ripping at carrion under the sated eyes of the lion that killed the beast. Behind it the dragline, which is powered by electricity, trails the umbilical cable that runs it and leads, inevitably, back to the power plant.

Today the dragline is visible up on a ridge. There have been sporadic snowstorms over the past week or two, so the progress the machine makes in the earth can be marked on the spoil piles by the depth of snow. What the machine has piled today seems almost black; its day-old debris is freckled gray, and a week back up the trench the spoil is pure white. The impression is that new mountains are boiling up from the earth, hot and basaltic at the start, and slowly cooling to marble.

Tonight I drive past the dragline in the dark, headed south to a meeting of the Rosebud Protective Association. The big machine works in the light of huge greenish spotlights on the brow of the operator's cab and on its boom. In that light the soil and rocks its scoop moves look pale and fleshless and already long dead.

This month's meeting of the RPA is being held at Wally McRae's home. He had invited me to attend. I am glad to; I'm curious to detect any sense of elation among the ranchers: District Judge James Battin is scheduled to rule soon on whether Three and Four have to meet the new Environmental Protection Agency standards or whether the plants were under construction before June 1975, and there are rumors all over the county that if he rules against the company the new power plant will die.

It is a hard, cold night. The men and women of the association sit around the central coffee table and back slowly into discussion. The room is redolent of the Old West tonight; the clean Levi's, the pointed boots, the snap-cuffed shirts. In a cabinet there are moccasins, feathers, shells, polished bone, and old silver. There is an old railway lantern on top of the cabinet, and a silver kerosene lamp without a chimney. The members drop their ashes into the spur-shaped ashtray. Among the ranchers here are Bill Gillin, sharp-faced with bristled gray hair; Don Bailey, deep in an easy chair, flannel shirt rolled up around thick arms; Nick Golder, leaning back with his hands behind his head and his right ankle on his left knee, eyeing the group with a faint gleam of amusement in his intense blue eyes; and Harold Sprague, tough and lumpy, with one ear missing. Sprague's position on the coal development is the most ironic: He doesn't have electricity.

In the background a coffeepot burbles and Ruth brings out cookies. The ranchers fill Styrofoam cups and eat cookies with restraint. They talk about what kind of overpasses they should urge be built into the road from Colstrip to Forsyth to allow cattle to pass; they discuss the upcoming legislative session and the likelihood that taxation and reclamation laws may be changed there. Bailey doodles on his agenda, stabbing it with his ball-point pen. Gillin takes notes on a yellow tablet on a clipboard, and at one point Golder leans back in his chair and looks out over everyone's heads and says:

"There's a good deal to be said for doing something that's *right,* because it carries some weight just because it's *right.*"

They talk about reclamation, complaining that the companies, though required to do so by the law, aren't putting the land back to its approximate original contour. And McRae points out that when the overburden is broken up and moved, it expands by about 30 percent.

"I wish," Sprague mutters, "that it would do that in a posthole."

There is talk about the funding for the association, and Gillin makes a suggestion that may be calculated to irritate the environmentalist members of the parent Northern Plains Resource Council, or maybe just to tease them. "Everyone," he says, "ought to donate one coyote hide to the RPA." No motion to the effect is passed. (When I ask McRae later if the idea ever blossomed, he says, "No; coyotes were worth too much this year.")

Then they talk about taxes and whether or not to advocate that electricity itself be taxed either as it leaves the state or as it leaves the power plant, and McRae twitches slightly with irritation.

"Here's our dilemma," he says finally. "It's the same as with the coal tax [Montana's 30 percent tax on mined coal]. If we're for a tax on electricity, it encourages development; if we oppose it, that encourages development." His theory is that, even though much of the coal tax money goes into a trust fund for a coal-less future, the rest becomes such a welcome source of revenue for most state agencies that they begin to look fondly upon the development itself and become predisposed to take action to promote it. In addition, he says, the tax, in the years since its passage, has been used more as an equalization tax than an impact tax, funding schools in places other than towns such as Colstrip where, the administrators say, capital improvements—the plant—now provide enough tax base to take care of needs. As the money is spent that way, McRae fears, it will be dribbled away, and when the time of worst impact hits—when the coal boom ebbs and the towns are left stranded on a beach of unproductive land and diminished resources, the money that would help them will be gone.

There is no hint of euphoria in the group tonight, not even a tickle of anticipation. It is as if these men and women have been shown victory too many times before and then had it yanked away. Instead of talking about the pending decision they discuss Montana Power Company's upcoming request to the Rosebud County Commissioners for industrial revenue bonds to finance the pollution control equipment for Colstrip Units Three and Four. Revenue bonds are tax-free bonds that can be issued by a government agency to assist industry in locating in its area and in meeting legal requirements such as pollution control laws. The bonds are guaranteed by the companies, not by the county. The county issued revenue bonds to finance construction of the pollution control equipment for Units One and Two, but this time the request is for far more bonds: a total of about $350 million. The ranchers are concerned about the bonds because they would save the companies about $80 million, and they see this as a subsidy of one industry—power generation—at the expense of another—agriculture. But the meeting trails off into reminiscences without any promise of action—Old Ambrose, he used to drive all the way to Forsyth in his old Ford tractor, once his license was taken away. Yeah, a mule once kicked him and got hurt so bad they had to shoot the mule. But there is a promise in the air: The bonds won't get through easily. And I wonder, sitting out on the edge of the meeting: Is this still another hurdle they are setting up for the power plant, before it has even crossed the one it faces now? If so, it doesn't look very high.

Afterward I talk to McRae late into the evening. He is in a somber mood. He had been arguing tactical dilemmas all night. He is sick of them. Now he reads me a poem he has written one similar night.

> The Land*
> Gestating in the mid-year
> Seldom did it show.
> Youthful green in springtime.
> White, and gaunt with snow.
>
> She who each year faithfully
> Gives her produce away
> Never asking compensation.
> Quiet, unassuming is her way.
>
> You'd ravish her with mindless lust,
> Then curse her for a whore.
> You've never loved her as I have
> Or you'd respect her more.
>
> Rip open her hard belly.
> Tear her vitals out.
> Sew her back with zippered tracks
> That wander roundabout.
>
> You'd prostitute her beauty
> With cosmetic care.
> Strip her of fertility
> And leave her prostrate there.

* Reprinted by permission from *It's Just Grass and Water,* copyright © 1979 by Wallace McRae.

"That's a little bit of Wally McRae," he says, with a hard laugh. And we talk on. Somehow we get to a pair of stories he remembers from the months of hearings on Colstrip Three and Four, up in Helena.

One evening after a day of hearings, he recalled, a small group of Northern Plains Resource Council members and staff were sitting around a table in Jorgenson's Restaurant in Helena, having dinner. The group, which included McRae and Kit Muller, a taciturn, slender staff member, was conversing quietly when one of the lawyers for Montana Power Company, Jack Bellingham, walked past.

"Say," Bellingham said, "how do you have the time to eat? I thought you guys spent twenty-four hours a day trying to figure out ways to stop Three and Four."

Remembering, McRae sits at the table, staring down into the empty kitchen, his face hard. He pauses in his story. Now his narrow eyes flick sideways at me and back to deadpan, and I am reminded again that somewhere in him McRae laughs and laughs at the terrible absurdity of this whole frowning matter. But he keeps his sense of farce in check. He continues:

"So Kit said: 'No, Jack,' he said, 'you'd be surprised at just how little time it actually takes to bring a major corporation to its knees.' "

There is a long silence. McRae chews on the soft end of a match. He is in a mood to reminisce, with the possibility of a favorable decision from Judge Battin nagging him with a hope he hates to acknowledge.

The second incident, too, happened at Jorgenson's Restaurant in Helena, a few weeks later. One of the members of the Board of Natural Resources had come up to McRae during a recess late in the hearings and suggested a compromise: The approval of one 350-megawatt plant at Colstrip instead of Three and Four. McRae had said his community had compromised enough with One and Two already, and later in the evening at the restaurant he and Muller argued it out.

"Well, we've got just one 350-megawatt plant and we've won," Muller said.

"Ohhhh, no we haven't," McRae said. "You know that a 350-megawatt plant is not a victory." But it was late, and the men were weary, and the hearings had dragged on for what seemed like decades while the future of Colstrip hung on the rim of decision like a basketball waiting to drop after the buzzer has sounded. And Muller sounded bitter, although McRae said later that Muller was just testing McRae's commitment.

"How much longer are you willing to fight this thing?" Muller said. "How many people are you willing to keep putting in the oven till it just burns them up?"

And McRae, with the almost fanatical determination that was there in the winter of 1964 but that was born in his family long before he was, said:

"Just enough people that it takes to win."

48

A consortium of power companies won at least a temporary victory over the Environmental Protection Agency Thursday when a federal judge gave the go-ahead for construction of Colstrip Three and Four. Judge James Battin, in a 33-page decision, ruled that the companies had begun construction of units Three and Four before June 1, 1975. . . . [The ruling] left the door open for future hearings.
— The Missoulian
January 28, 1977

On the wall of Martin White's office in the new Western Energy building in Colstrip hang the heads of three animals: a wild ram, a wolf, and something that looks like a large deer. When I arrive at 8:00 A.M. on a cold January morning, White stands up behind his desk, short, athletic, with broad shoulders, friendly. "Come on in!" The sheep looks helpless, the wolf snarls, and the other thing blows cold steam from putty nostrils. I pat it on the snout, shake White's hand, guess, and say "Nice elk." "Caribou," White says. "Got it in the summer of '74. Something I'd wanted to do all my life."

White has already been for a two-mile run and played a game of handball. His face is ruddy with the exercise and the outside cold. This morning the power plant sends a huge white cloud straight up into still air, a cloud which dissipates two thousand feet up. The town lies crisp and clean under new snow. There are trucks out on the streets scattering coal gravel on the roads.

White is ebullient today. "Did you see that last night they announced Judge Battin had overturned the EPA? He's given grandfather status to Three and Four! It's not going to be black and white, I'm confident. But I'm sure Judge Battin spent a lot of time sorting through the legal complexities."

With that as a start, we talk about the power plant.

"I don't like to take a stand on something unless I believe in it myself. I just really don't think that I can do that. And I've told Paul [Schmechel] and George [O'Connor] that. But I do feel that Three and Four will be a benefit to this area. . . . I'm not enough of a chemist or engineer to know whether the plants are polluting or not. I have to believe our engineers. And when they tell me that they're capable of doing something I just have to believe them. And I know most of those guys and I just don't think that they would lie to me."

White is as sincere and optimistic as Wally McRae is sometimes cynical and depressed. There is none of McRae's introspective tendency in White at all, it seems. They are so different and their positions are so representative of what is happening here that they are often picked by television crews as symbols of the conflict. "Martin's a perfect foil for Wally," Patricia McRae said once. "Martin comes across as a blond, sunny, all-American boy and Wally comes on dark and gaunt and sinister." Perhaps a friendship across the lines was inevitable. "Sometimes," Duke McRae said, "I think they get along a little bit too good. I think they get a hassling at each other until they learn a few too many trade secrets and, you know . . . I think in some ways they both half enjoy it."

White's acquaintance with McRae began back in 1974; it grew to friendship when both men were named as directors of the Colstrip bank, and it was continued by little things. White, for instance, got manure for his lawn from McRae's corrals. "Martin and I have the best yard in Colstrip," McRae said once. It was a friendship probably built more on antagonism than on common interests, begun partly as a demonstration of both men's belief that conflicts of politics shouldn't descend to politics of personality and partly as an expression of hope.

It reminded White of his talks with his father.

"Dad and I could discuss very intensely, and know that was all we were doing," White says. "There's not many people you can do that with; not many people understand that all you are doing is protecting or developing your side of the issue. That's one thing Wally and I can do; we can sit and argue and recognize that all we're doing is presenting our viewpoint and I'm not trying to get after him and he's not getting after me or anything."

To McRae some of the friendship was based on candor.

"Martin has never lied to me," he told me once. "There aren't many people that work for the Montana Power Company that haven't lied to me. He never has. And even when it would have been much easier for Martin to lie to me, he hasn't."

Today White shows me a copy of a paper he recently wrote with some advice from McRae for the Atomic Industrial Forum, Inc., and the Edison Electric Institute. Its title is: "A Report on Mitigation and Monitoring of Socio-Economic Impact of Large Scale Energy Development Projects." Its existence is an indication of White's success at Colstrip, of the extent to which he has been able to achieve his vision: that Colstrip should become the best living environment for coal workers in the West.

When White talks about this vision, he becomes as excited as he does when he talks about his father. He slings his concepts out from Colstrip in arcs of theory, hoping to apply them as principle to all such towns. The idea is simple and seems obvious: A worker who has a pleasant living environment produces more than a worker who lives in organized squalor; and therefore expenditures to improve that environment, which show no direct return, do pay off.

"I really think you've got to be concerned about people every minute of their day," he says. "Let's suppose we just say to hell with the workers, let 'em buy tents. And somebody'd come in and build a trailer park, and they'd do it as cheaply as they possibly could because they'd worry about when they'd get their money out. And it'd be a junky-looking town, there's no doubt. And the quality worker would not come here. He'd say I'm not going to take my family down to a place like that. So the town would look junky, their streets wouldn't be paved, and when they went out to the mine they wouldn't develop pride in things like keeping the dragline clean. And the productivity, the amount of maintenance would all be reflected in that.

"And you know I can't put a dollar sign on that. I can't tell our management that last year we saved you two million dollars because the town of Colstrip is as it is instead of being a trailer camp. But I know this is the case." He thumps on the manuscript before him on the desk. "And this whole paper is really to try to convince the whole utility industry that that's the case."

Although Wally McRae complains sourly that White's paper is nothing but a treatise in how to obtain government funding to do things to a community that the companies ought to do on their own, White has indeed changed Colstrip from the town of mud and depression that he moved to in 1974. Today, working from plans designed by a town planning consultant, White has created a town which, with the exception of the Burtco trailer court scar, looks like any reasonably well-planned suburb. There are winding streets lined with neat homes that look different enough one from the other to belie one-company ownership. There are three tennis courts, an undulating park with a wading pool in which stand concrete elephants and whales. There is the community center with its gymnasium and swimming pool. There are softball diamonds and tot lots. The surge pond, which furnishes water for Units One and Two, has been stocked with pike. A community shopping mall has been built, though not completely filled with businesses, and a small supermarket with competitive prices has replaced the store that Lois Olmstead called the Pluck Me. In four years the community has changed from the metal ghetto of its early fame to a town which the *Christian Science Monitor* singled out in comparison to other coal boom towns as "an example of how companies involved can ease the adverse effects of rapid development." The only things that make Colstrip vitally and permanently different from any other pleasant small community are that it is an island surrounded by a sea of mined and unmined land that changes even its shape and character under the wind of energy demand; that it is dominated by the power plant, whose voice reaches into every home; and that, any day now, it may boom and all the serenity will be inundated again.

But White says he is ready for it this time. When he looks down the barrel of Three and Four he sees a town of about five thousand, "just big enough," he says, "to support one store for each kind of business." And this time, he says, he's going to prepare for it.

"I would like to have the construction finished on the town when the people come," he says. "That's hard because you've got to squeeze those dollars out of [the company] and you know they fight with you until it's a crisis down here. But I've just started really with no authority. I've made this assumption: They've assigned me to handle this. They think I did a good job on One and Two, apparently. They're busy men, so they don't want to know about details. They just want me to get it done." He pauses, the excitement in his eyes again. He can't resist adventure, even if it means going out on a limb in his job.

"Those are the parameters I'm working under right now," he continues. "And so far nobody's refused to pay the bills. Everything's going to be paved—there won't be a Burtco this time. I hope the town will be a collection of small communities because of the hills. We're going to have tot lots shot through the area. I hope to rig it so we get about a ten-foot-wide bicycle path that ties each of these together. We're going to have five-star trailer courts—the best in the land." He grins. "I wrote 'em a letter and said it's going to cost you twenty-eight million dollars and no one's written back and told me not to spend it. So I'm just working as though I had full authority to go ahead."

White overflows with optimism. Sometimes he reminds me of one of those big balloon clowns, weighted at the bottom, which bob over when hit and then roll back up, painted smiles ready for more. He's been bludgeoned here and there, and he's still standing, reaching with delight for the next assault. His ballast is either blind hope or courage, and I prefer to think it's the latter.

White sits at his desk, idly cleaning his fingernails with a silver letter opener. When he is passionate about an idea, sometimes he talks loudly, but often he lets his voice go quiet. We have been talking about coal and the power plant, about why he is so determined to make Colstrip a community rather than a camp. A bit unfairly, I ask him about the state coal tax, the whopping 30 percent bite that the companies have often attacked as diminishing their competitiveness with Wyoming coal. My question is kind of a test of him—see if he toes the company line. Does he disagree with the tax, whose intent is to provide for the coal-less future?

He pauses at the question, as if it were a snake in the road. Then his voice gets very quiet and a little tense. I can suddenly hear my tape recorder running, grinding out little curls of recorded time.

"Maybe not 30 percent," he says finally, slowly. "But I don't agree with our people resisting the tax." He pauses, walks slowly around behind his desk and looks at a photo on the wall. It is a picture of Butte before the Berkeley pit ate half the town. And he ties it together, the tax, his determination to make Colstrip livable.

"I looked at the Anaconda Company," he says, still very quietly. "And I thought—look at all they've taken out of the ground and what they've left for the state . . . and all they've left is a hole in the ground and a town that is dying."

I leave White and walk down to my trailer, across the tracks and up past the power plant, passing by so close that I can feel the tiny prickles of frozen mist that drift down out of the steam of the cooling towers. There is a smell of sulfur in the air here, and for a while I thought it was the plant; ironically, the smell comes from a hot-water well tainted with sulfur in the bed of Armell's Creek. Walking past the plant, engulfed in its noise, I think about something Wally McRae told me one day when I visited his home.

On some weekends in the fall, he said, when the pressure in Colstrip was intense, White would drive down and go out hunting on McRae's land. And he would come in with the sun dying in red and amber on the hills and the dry grass turning gold and he would grin at McRae in the ranch house yard.

"Sure appreciate you allowing me to come out here and get away," he'd say. "So much of the time I've got my nose to the grindstone that I don't appreciate really what's around here."

And McRae would flick his narrowed eyes over at White and glance away, a little wickedly, like a man teasing the edge of his knife with his thumb.

"You know, Martin," he'd reply, "you shouldn't come out here. It's too much of a stress you're putting yourself into. 'Cause we're doing numbers on you constantly. And the only thing I know for sure about you is that when you go to bed at night and the light's out and you don't have to hold back with facial expression or anything else, the last thing you say to yourself when you go to sleep is, 'Damnit, they're right.' "

White would laugh, and they'd go into the house with the steer-hide on the couch and the lupine glowing from the painting on the wall, drink coffee, and talk about other things; then White would go home, back to his vision of Colstrip to be—the friendly town on the prairie. And McRae would remain on the Rosebud, just out of reach of the voice of the power plant, and think of Colstrip as it was—a military base, with its rootless enlisted men, its war widows, its animosity with its neighborhood, and its lonely but dedicated commander in chief.

"I'll give Martin credit," McRae said. "He has tried every way a human being possibly can to make chicken salad out of chicken shit."

49

Presidents of five Northwest utility companies will be meeting in Butte this week to decide their next play in the struggle to construct Colstrip Units Three and Four.

—*Forsyth Independent*
February 3, 1977

Teased by the edge of a shadow of steam, the sunlight in Vic Jungers trailer goes bright and dim, bright and dim, as if someone had attached a rheostat to the universe and was playing with the sun. The wind is from the south again. In the tiny yard the five Doberman pinschers are restless; they trot back and forth in lean irritation under the flashing sky.

Jungers talks on and on. The cuffs of his denim jacket flap loosely around his wrists. His black hair is unkempt; he runs his hand through it. There are black-and-white photos of churches and graveyards on the wall. A redwood bole coffee table stands on reinforcing bar legs in front of the couch, bearing a copy of a book by William Faulkner. Jungers thumps his chest with his two middle fingers. A silver tooth flickers in his mouth.

"The ranchers basically see the end of a lifestyle," he says. "I see the beginning."

Jungers, playing in his darkroom, has made a curious photograph. He took a picture of Colstrip Units One and Two side by side, enlarged it, turned the negative over and put a second image, almost double the size, on the paper next to the original shot. And there, when the print swam into life in the chemical bath, was the future: Three and Four standing huge and symmetrical beside their smaller brethren. He

214

made copies of the print and gave them to other ironworkers. He hung one on the bulletin board at the Bechtel office near the time clock, where the men look at it with a kind of smirk, as if, seeing it, they know something the opposition doesn't: that Three and Four are inevitable.

Jungers was part of the first wave of construction workers; part of the first sign of the fundamental change. Jungers came as a boomer, a term he prefers to "transient."

"Boomers are a particular type of construction people that just move from job to job," he says. "Only stay a few months and move on. And when you have a job starting in an area like this and there's not enough manpower, you get a lot of these people. The boomer just comes, works, and if it is a good job he'll stay. If it isn't a good job he'll just pack his bags and go to the next one, till he finds one he likes. Generally they're a much rowdier type of person; they live a lot harder life, a lot faster life."

The epitome of freedom. Endless travel, new towns, travel again. Jungers is sardonic, without attachments. "I have maybe five friends. The rest are acquaintances." He's divorced; he's been living with a woman for three years. He likes to backpack in the Absarokas, a mountain range near Yellowstone; the subtle beauty of eastern Montana is all desert to him. Perhaps he shared in the joke that ran through the town early in the construction boom, that Custer and the Indians had fought the battle of the Little Bighorn to see who had to stay.

Not long before Jungers came to Colstrip he was washing ducks of oil spilled in San Francisco Bay and picketing the offices of Standard Oil in protest. So when he arrived the globules of environmentalism still clung to his feathers. Early in his working days here, while he was helping raise the structure of Units One and Two, a television news team from Texas interviewed him, and he told the reporter he thought that "along with the energy crisis, coal development was a rip-off," then, characteristically relishing the threatened position his remark put him in, added that his frankness would probably cost him his job.

But the show never aired near Colstrip, so there was no repercussion, and Jungers remained safely unknown. But by then his mood was changing anyway.

"I came to the realization that if I was going to work iron, I was going to be involved in power complexes and whether they built them in Montana or not was irrelevant to me," he says. "I felt the state needed some more industrialization."

Jungers' voice is a curious mixture of gravel and honey. It is languid and rough, and the ideas roll out without effort, either because he has previously thought them out at length or because he doesn't care. Or perhaps one disguises the other, the way the languor in his voice covers its power. And as I listen—questions are seldom needed; I just provide an ear—I wonder if maybe the weariness and the sardonic tone cover another kind of longing: for complete submersion in an issue, the kind of dedication without question that is easy for a less complex man but almost impossible for him.

When Jungers first came to Colstrip, the plant was just a job for him, not an issue. At last he had money. In his most recent year he made thirty thousand dollars. What does he do with it? "I indulge myself. I haven't saved much. I have about three thousand dollars in camera stuff. I go on vacations to Las Vegas, San Francisco, spend two, three thousand. After seven and a half years of college I got tired of eating beans."

But the controversy grew on him like an old itch.

He had always suspected that the environment movement was engineered by the Nixon Administration as a device to make people forget about Vietnam, so in spite of the oily ducks his loyalty to it was slim, tied only by a thread of continuity that the same liberals who fought for blacks and peace in the 1960s were the ones planting gardens and picketing nuclear power plants now. So it was not hard to slip away, particularly when an even older loyalty, trade unionism, beckoned to him from the other side.

His conversion began in Forsyth and was completed one day at a meeting in Missoula, not much later.

It was one of the interminable series of meetings on Units Three and Four, and Missoula was a college town. For the students there was no Vietnam, precious few blacks; all that was left was Colstrip. So the auditorium was packed and the crowd restless.

"I expected them to be really out of line," Jungers says. "There was tension."

The first few speakers were received politely. A professor arguing about the emissions problems. A student who, Jungers thought, was making a long joke about telling how many bees would be put out of business by the power plants and their associated coal mining, but turned out to have been serious all along. When a speaker representing Portland General Electric promoted the plant there was a hissing in the room. Then a pipefitter got up on the stage. Jungers knew him: He was a big man, middle-aged, a poor speaker. It took effort for him to even be there.

"My name is Leonard Shaeffer," he said. "I'm a pipefitter. I'm employed at Colstrip."

The noise of the audience rumbled up and crashed on him. Boos, laughter, shouting.

Shaeffer tried again. "And I'm in favor of Colstrip Three and Four." The noise continued.

"He just went red from the neck right on up," Jungers says. "And I'm sitting right down on the side of the stage, staring straight up at him, toward center stage. And I felt sorry for him. He tried to maintain his composure. And he tried once or twice to speak again and they booed him every time he tried to say anything. Oh, God, I knew he wanted to jump out in the audience and just bang some heads, and he's a big man."

But Shaeffer just stood there and waited. The noise slowly died.

"Thank you," he said slowly. "For your attention and courtesy." And he walked off the stage.

"And I thought that was pretty damn cool," Jungers said. "I don't think I'd have said something like that; I would have been pissed and told them what I thought of them. And I felt bad about it and I was angry, and I was kind of embarrassed at having been part of the academic community."

So Jungers closed the door on that side of his life. He finally became an alien to that world which he had so spent himself to join, had so fully sacrificed to live in for seven and a half years, and had so longed to be a part of for life. The cool courage of the pipefitter under stress, like the Hemingway ideal, caught him, moved him, and cemented his position. Even if, in his endless contemplation, he knew he could never give himself fully to a cause, at least he knew from which side he started and to which side he would give any loyalty he might discover in the wanderings of his mind.

It was an odd position, aligned with Montana Power Company, even if the unions involved in the construction were on that side, too. The company did not impress him, either in its candor or its methods of construction. Montana Power, he says, used cheap equipment in invisible places where expensive equipment was necessary, and used expensive equipment in visible places where it was just window dressing. And his socialistic beliefs were undiminished by his shifting of allegiance from the collegiate brand of liberalism to support of old-fashioned progress. Government ownership of utilities, he says, would be preferable to this rush for profit. And he begins to ramble off into a kind of long-winded rhetoric that itself could make you nostalgic for those past days of ideological zeal, when the words had the weight of defiance, and you thought there was enough power in the crowd to change the world.

"Communications is going to do it for us. Next hundred years there'll be so much change. It's inevitable. We'll make the world a smaller place and people will become more interdependent and less jingoistic about being American and more international in outlook. And then I think the time would be right to have communication and all of this run by the government instead of private enterprise. I am for nationalization of all utilities, communications, things like this. One thing, you know, during the war Mussolini made the trains in Italy run on time, probably the only time in history. Like the man said on '60 Minutes,' the government could have everything if they could make the streets as safe here as they are in Moscow. I'd let 'em look through my garbage."

While we talk, Jungers is packing. He's leaving for Canada tomorrow. He'll be working on a Canadian project to extract oil from tar sands—once again, energy. The more he studies it, the more he thinks the world's demand for energy may create the holocaust he sees coming. "Right now, if we were cut off from our oil supply—what do you think we'd do? I think we would go to war. I would like you to think about that."

In his briefcase Jungers is stacking books, his most precious baggage: *The Confessions of Nat Turner*; the *New Enlarged Pocket Anthology of Robert Frost's Poems*; Richard Farina's *Been Down So Long It Looks Like Up to Me*; *The Rape of the Great Plains*, by K. Ross Toole; *Spotted Horses/ Old Man/ The Bear*, by William Faulkner; *Trilogy*, by Diane Wakoski; *The Pipefitters Handbook*; *Words*, poems by Robert Creeley.

Jungers is animated. The trip excites him. New places, new people, new work. The boomer in him has been dormant for so long, stuck here in Colstrip for six years, that he had almost forgotten the feel of it. He's leaving his trailer, the Dobermans, and Connie, here in Colstrip; the job's a short one—six or eight weeks—so he'll be back.

Now that he's going to be free, for a while, from the spell of Colstrip, he talks frankly.

Jungers has, perhaps, the worst combination of perceptions for peace of mind: cynicism and idealism. In that he is a bit like Wally

McRae. He sees the world incredibly warted and perverted by imperfect beings, but he cannot relinquish his high expectations. And one of these expectations, which is fundamental to his belief in the value of trade unionism, is that union people have pride in their craft: that when they demand high payment for their work they follow up with performance. That was not what he saw in Colstrip.

"There does not seem to be the camaraderie or the pride there used to be in the unions," he says, "and that has a lot to do with their outlook on how they work, how they perform."

Out in the plant, during the days of peak construction, the air would be sweet with the smell of smoked marijuana, Jungers says, and the mood would be mellow and easygoing. There would be half-hour coffee breaks, long conversations, and delays to force overtime.

"People were telling people, 'Don't make any walls until after 10:30, so we'll be sure to get two hours' overtime tonight.' Really! That has happened! Cost overrides have been tremendous.

"Used to be," he continues, "when my dad was figuring jobs and stuff for a company he worked for, they figured like six hours' work for eight hours' pay. I imagine if they got four hours' work out here they would have been happy in a lot of instances."

How do you get your own men to work in such an atmosphere? Jungers, as a foreman, tried. He'd sit a man down who'd taken a long break. "Look," he'd say, articulating his own fear. "The union is your best friend. Your mother, your father, your rich uncle, your shield. If it wasn't for the union you'd be riding a horse instead of that new four-wheel drive you've got. But you start dropping off on production, along with everybody else, and before you can refill your coffee cup the company's hired some nonunion contractor and you're out in the street. You don't know what it's like to look out your window and watch them build something that's nonunion. Yet. But it's happening. The union movement's not dead, but it's dying rapidly from paralysis. And when you sit there and take a thirty-minute coffee break every chance you get, you're a part of that."

"And they'll sit there and agree with you and they'll blame it on another craft," Jungers says. " 'Yeah,' they'll say, 'look at those rotten bastards over there. They're just sittin' around. They just sit around and eat lunch.' You'll walk away and there he is sittin' there eatin' lunch, too! Ten o'clock. What do you do? You fire him. You get more just like him because Colstrip's not attractive."

And each new man on the job, each turnover, increases the risk of accident. Some men cleaning a water pipe were nearly drowned when someone turned on the water by mistake. Two men working on a scrubber fell and barely escaped with their lives. On Jungers' crew a man lost a finger.

"An avoidable accident," he says. "Most accidents are human accidents. But they are lucky out there. Really lucky. My God, they were so fortunate out there it's unbelievable. Every day during the peak period I waited for somebody to get killed."

After the Forsyth meeting, Jungers joined the hearings circuit. He, like McRae and Bailey and eventually Duke McRae, went on tour. Mike Moon, running the series of public meetings, became accustomed to his large, sharp-edged face and his sweet-gravel voice: "You people, the ranchers who controlled the state for so many years—you're the ones that can change the laws. If the air pollution laws are not stringent enough, you have no one to blame but yourselves. Why worry about it? If the plants don't meet the standard, the law says they can shut them down. Think about that.

"Historically, Montana has been more heavily dependent upon agriculture than other states. Therefore, any change in agriculture had a greater economic impact. But Montana only accounted for 1 percent of all cash receipts from farm marketings in the United States in 1969. The present role of agriculture is declining and so is its importance. Coal in the future will become Montana's largest industry. Think about that."

It was all spoken with a sort of half-sad, half-amused world-weariness that made him sound either like a wise outsider with advice for those grubbing in the dirt or a pompous college student, depending on which side you were on.

Jungers spoke at Missoula, Polson, and St. Ignatius. Once he took a week off to attend two meetings far to the west of Colstrip, near a town in which his brother owns a bar. He spoke, then visited. The visit was probably more important than the speech. But the surprise came when he returned, and found that he had been paid full wages for the forty hours he had missed, plus mileage.

"When I got to the meeting here were all these guys from the plant, and they were pretty sure they were going to be paid. I saw Myron Brien there; I was flabbergasted. I think some of them would have been pissed off if they hadn't been paid, really."

I ask him how he can justify being paid to speak at meetings designed to gather unfettered public opinion. He admits that the payment allowed opponents of Three and Four to label him and the other workers as lackeys for Montana Power Company, a reminiscent word.

"It does detract from the credibility of whatever you say," he says. "It's a very disparaging situation. But I felt very strongly about what I did and about why I did it."

But then he gets defensive. Besides, he says, "when you make four hundred dollars a week it's kind of hard to go someplace and do something like that free. If I were a rancher and most of my work was in the fall and in the spring or in the summer and my winters were free, I wouldn't be out anything. So what the hell's the big stigma?" Jungers, his cuffs flapping, thumps on his chest again with his middle fingers. "And as far as ethics go, taking the money, I don't think that had anything to do with what I had to say."

At thirty-seven, Jungers is a brooding man. He feels his age. He'd still like to be a connector—the man literally at the top of the ironworker's profession, but he feels he's too old. "You get signals," he says. "Your body's trying to tell you to slow down. They irritate me."

And he still has his ambitions. Ironworking has been his door to financial security, but he left it once before to try to learn to teach, and now he's back. In his endless contemplation, in his consciousness of class that he hates but recognizes too clearly to be unaffected by it, he must occasionally remember that description of an ironworker offered him by his ex-wife—that an ironworker is a fellow with a suitcase full of dirty clothes and a hard-on. She was the one with ambition, he says. "I just wanted to be happy, though I guess that's ambitious—I've never really, really been happy." And maybe the description continues to nag him, like any half-truth that you wish was all lie, because his ambitions are not as modest as he says.

"When it comes right down to it," he says, "what is my job? It's pretty insignificant when you consider all mankind."

That's an easy way for anyone to get depressed, I say.

"It is, really," Jungers says. "I think about it all the time. I've always wanted, I've always wanted to be a great person in some way or another, and I realize I probably never will."

There are realms of greatness that he wants to explore. The great American novel. The great book on the American literature of the 1930s. They are all there, waiting, unaccomplished by others. Why not try them? I ask. His mind brims with the need for expression, stifled in this trailer, in this town, hemmed in by the walls of spoil, drowned by the noise of the plant. Why not at least make the attempt?

"You know why I don't," he answers. "This is the reason I don't do a lot of things." His jaw juts; the dogs are barking outside; the light in the trailer has been gone a long time behind the spreading column of steam. Suddenly I wonder: Is he trapped here in the shadow of the plant, or is he hiding?

"This is one of the *keys* to my life," he says. "If I wrote and it wasn't any good, I'd never write again."

50

Joseph McElwain [recently named president of Montana Power Company] told members of the Forsyth Chamber of Commerce that bureaucratic gobbledygook and legal questions are holding up construction of Colstrip Three and Four. "We have been told by EPA that they would not issue a permit until the determination of air standards on the Cheyenne Reservation is made," he said. . . .

—*Forsyth Independent*
March 24, 1977

Today I awake to an earthquake, or so it seems. I am lying half asleep when there is a distant, brisk rumble, and the trailer shudders. Sometimes my little wheeled house moves in just the wind, creaking and swaying to the gusts, but this is more of a jolt, a sudden passing tremble. I have lived in California; with the first shake of it I leap out of my sleep entirely and am at the door, in time to see the black cloud rising above the mine and to remember, sheepishly, that this happens every day, often more than once. It is the blasting at the mine, the daily shock that jolts acres of overburden or coal into rubble in one swift crack of nitrates and oil. You get so accustomed to the little booms and shocks that catch you on the street or shopping or in the Colstrip community center gym that you stop noticing them until they surprise you in sleep with sudden images of a slipping earth.

Today, Myron Brien. Enigmatic man. But then, who isn't? Martin White, with his straightforward eyes and sincerity?

I go to see Brien in his new fourteen-foot-wide mobile home over in the permanent trailer park, south of town. This park, unlike the place I live, has lawns, curving paved streets, and quaint black lamp posts.

223

Brien is on day shift, so I have to wait until evening. I enter to the sound of dice rattling on a table; Brien and his wife and one son—about twenty—are playing Yahtzee on a table in the spacious living room while a younger child watches television. The trailer is a recent acquisition; it is done in green and brown Mediterranean. Brien, true to form, has a Bull Durham tag hanging out of his shirt pocket.

Brien drops out of the game, offers me a drink, then mixes himself a screwdriver. He is tall and lanky. He gets around slowly, but not wearily, as if he has just been lubed but the grease in his joints is thick.

You testified at a number of meetings, didn't you? I ask.

"Quite a number of 'em, yeah. Hell, I still think they need a place to generate electricity, and coal mining goes hand and hand with it. That ensures a place for employment. You don't have to depend on the market as such. You've got generators One and Two and possibly Three and Four. That ensures a place for this coal to be used. If Three and Four go in, it'll make my job a little more secure; it'll make jobs available for other people in this area like myself."

Brien's voice is deep, his talk is abrupt. He chops his meaning out of the language the way he might build a road with a bulldozer, pushing and grinding colloquial stones into a rough but efficient statement. He is not so much dedicated to the advancement of the power plant as he is scornful of its opponents.

"They say they'd like to go back to the old days," he says, echoing some of the testimony he made two years ago at several of the meetings. "Well, they don't even know what the hell the old days were. When I lived in Lame Deer, and Busby, you packed water from the spring, you'd go to the outhouse. Hell, we had to chop wood every night, pack ashes. Hell, if *they* had to chop wood they wouldn't know what end of the ax to take hold of."

Brien was described by one rancher as Montana Power Company's token Indian during the meeting and hearing phase of the debate over the power plant. Each time he spoke he announced himself as an enrolled member of the Northern Cheyenne tribe, and although at least once he felt obliged to mention that he did not speak for the tribe, the relationship had force. It eroded the strength of the other Cheyenne speakers, who sought to give an impression of solidarity, of a tribe united against a threatening outside force.

But Brien does not repudiate his ancestry. "You know a lot of people cuss Indians," he says, "but you will find that the Indians will

take care of one another, even if it means sharin' what they've got."
And I remember his Forsyth testimony, in which he referred with some
pride to the new power the Northern Cheyenne people were gaining
over the "Great White Father back in D.C." This could only have been
a reference to the coal leases the Northern Cheyennes had wrestled
away from the companies that had taken them earlier, and I wondered
about it when I heard his tape. And I wonder today how he feels about
what is happening on the reservation now. The centennial of the Cheyennes' long walk back from Oklahoma is just over a year away, and the
Indians are stirring again.

Almost two months ago I went to a meeting called in the Lame
Deer tribal center gym. The roads had been bad; it was a warm late-
winter day with snow and light sleet. Although no other members of
the press were apparent in the audience, people in the gym seemed to
be conscious of being present at the making of history, a mood height-
ened when the students from the Lame Deer High School filed quietly
into the room and up into the wooden bleachers. The day had been a
long succession of speakers, both whites and Indians, speaking in English or Cheyenne, all pleading, with numbers, scientific data or just
eloquence, that the purity of the Northern Cheyenne Reservation's air
be maintained. The occasion was a required hearing on the tribal
council's petition to the Environmental Protection Agency that the reservation air be officially recognized as Class I air. It was the beginning
of the education of the Northern Cheyenne nation in the politics of
clean air.

Clean air first became an important part of the legal structure of
this country in 1970 when Congress passed the Clean Air Act, an amplification of earlier attempts to control air pollution. At that time the
infant Environmental Protection Agency interpreted the act to mean
that general limits—national ambient air standards—were placed on
all air, no matter how clean or dirty it was. You could pollute to those
limits and then you were shut down. In 1972 the Sierra Club sued the
EPA, arguing that the legislative history and the wording of the act required not only the upgrading of clogged air but also the preservation
of unsullied air. The district court decision upholding this concept
reached the U.S. Supreme Court, where it was again upheld in a four-
to-four vote. The EPA was mandated, in the court's words, to "prevent
significant deterioration of air quality" in clean air areas.

From that decision came rules for the prevention of significant deterioration, which Montana Power was already worried about. These rules, which went into effect in June 1975, included a tiered definition of the word *significant,* allowing three different levels of deterioration, by area. Class III areas were places where "deterioration up to the national standards would be considered insignificant"; Class II zones were areas "in which deterioration normally accompanying moderate, well-controlled growth would be considered insignificant"; and in Class I areas, "practically any change of air quality would be considered significant." The 1975 PSD rules labeled the whole nation Class II and said that states, Indian tribes, and federal land managers could request redesignation of specific regions into other classes.

Since the creation of the PSD rules Class I had become a symbol of purity. A Class I area, conservationists had come to believe, would be a reservoir of the best air in the United States, a place of vistas unimpeded by soot or artificial haze, a place where you would want to breathe deeply and long before returning to the urban murk of Class III. Class I was the ultimate, the jewel of clarity in the box of stones. So far not one cubic foot of air in the United States has been called Class I. The Northern Cheyennes intend to be the first.

This spring, after the meeting in Lame Deer, the tribe published a volume called the *Northern Cheyenne Air Quality Redesignation Report and Request.* In it they gave formal support for their request for Class I air from several angles: health—respiratory ailments significantly shorten Cheyenne lifetimes; vegetation—increased levels of sulfur dioxide even at low concentrations can damage Ponderosa pine, the dominant tree of the reservation; and, indirectly, social impacts on the reservation caused by the construction and operation of nearby polluting industries. But it always came back to history and tradition. The long walk.

"In order to understand what this petition for Class I air quality status means to the Northern Cheyenne people, it must be understood . . . in the same context as the tribe's walk against impossible odds and almost certain death a hundred years ago to reach its homeland, its consistent determination to maintain the integrity of the reservation, and its current refusal of instant riches for the sale of rights to violate its integrity. Redesignation to Class I is in the same spirit as everything else the tribe has done during the last hundred years to secure its freedom and relative autonomy, and to retain the value and viability of its cultural identity. . . .

"All Northern Cheyennes want progress—a better life for ourselves and for our children. But we do not want the kind of a false progress that sometimes goes under the misnomer of 'economic development.' Economic development brought by outsiders to Indian tribes is often a false progress that strips them of priceless natural resources, disrupts their lives and traditions, and leaves them only dollars which are quickly gone. We want our own kind of progress that will work for us, not someone else's progress that will export our resources and leave us the consequences."

Although portions of the report indicated that it was not the intent of the request to stop development outside the reservation, it did make it clear that if the request were approved it would cast doubt on the future of Colstrip Units Three and Four. The report quoted an air pollution meteorologist for the Montana State Department of Health and Environmental Sciences, who modeled the possible effects of Three and Four and found that "the maximum concentration of [sulfur dioxide] expected from Colstrip Units Three and Four, and other proposed power plants in the area, would exceed the Class I increment on the reservation. . . ." Although the report tiptoed around a direct statement, the assumption among Indians and ranchers was that if the redesignation was approved Three and Four would be seriously threatened, if not doomed. The first officially recognized refuge of pure air in the nation would be something of a joke if the government subsequently allowed a major polluter to locate just upwind.

Today, as I talk to Brien, his native tribe is going steadily through the steps required to get its air recognized as pure. But Brien remains on the other side. The tribe no longer has his allegiance. Colstrip is his home.

"We're happy here," he says. "We can go into Billings and have a lobster dinner now. Six years ago it was a rare occasion we could go out and get hamburgers. One of the neat things about living in Colstrip is you can go away. We go play tourist in the Black Hills; we go to all the tourist traps there."

The Black Hills were the Cheyennes' holy mountains. Brien goes there to see the Reptile Gardens and the faces on Mount Rushmore, and to visit the bars in Rapid City.

The dice rattle on the table. Brien sips his drink. The television jabbers in the background. Outside the power plant roars.

"I'm making my own living," Brien says. "I'll do whatever I want being as I'm paying for it myself. They don't give me blue stamps and all that crap. Everything that comes across this table is paid for not by the government or the tribe or welfare. I don't know if you understand the politics of the reservation. It's a pretty goddamn vicious ball game. I just stay away from there. That way I don't owe them nothing and they don't owe me nothing."

51

... *Other hurdles include a court appeal of the Board of Natural Resources endorsement of the project last summer. April 4 is the deadline for the Northern Cheyenne Tribe and the Northern Plains Resource Council to file their briefs. The companies have until May 4 to reply.*

—*Forsyth Independent*
March 24, 1977

In the cold air the plant's great column of steam is always part of the landscape, ever changing.

Yesterday the two columns from the stacks rose straight, their bases tight and boiling, climbing and expanding to a wider cloud and directly from there up into a sedate bundle of white which hit a level probably a thousand feet above the stacks and abruptly vanished, as if the clouds were pillars on which a warmer air stood. The twin sheets of cloud from the cooling towers, however, seemed to blend quickly into one large one, but instead of rising perfectly vertically, they curved slightly on their way to the same vanishing point, bending in toward the columns from the stacks, as if in affinity. So the three lifting, always lifting pieces of cloud became a pair of soaring exclamation points and a swept question mark in the still air.

From almost anywhere within twenty miles of the plant the steam is visible. From over in the new side of town, with its winding roads, playgrounds, tennis courts, and ranch-style homes, it is most dramatic: The cooling towers stand between the homes and the plant; their steam almost completely hides the main structure; the plant appears only occasionally through windbreaks or eddies, so the town seems to be just any other suburb in which has suddenly risen a vast and miraculous cloud, filled, in the night, with the lightning of the strobes.

229

From farther away, up the road to Forsyth, the cloud becomes less distinguished, but it is always there, a small vertical twist of steam in a vista dominated by the horizontal lines of the plateaus and the sky, and the bases of any clouds. From a distance the column seldom seems to be straight; it curls its way up, as if its core was a vortex. Occasionally, when the weather is perfect—the air is cold and the humidity is balanced on the fine edge where temperature and dew point converge—the plant's steam never disappears; it just grows and spreads into a layer of low stratus that reaches down to the Rosebud and beyond. At these times the air itself is unstable, and low clouds often form on their own; as they roll in and congeal around this man-made layer, it seems as if the plant itself is the god of climate, a geyser of rainy weather.

Tonight the wind blows, and the scene is changed. Once again it is from the south, fifteen miles an hour on the ground, perhaps twenty at the top of the stacks. The cloud comes right over my head. The wind is remarkably steady tonight; the vortices made by the intrusion of the stacks into this stream causes a spinning turbulence; each of the two columns from the stacks is itself divided into two strands, each twisting inward toward the other, expanding as they rush north. The cooling towers beside the stacks produce a single sheet, unusually stable tonight, which sweeps overhead parallel to the clouds from the stack.

Tonight it has never grown dark. The trailer court is flooded with pale light; dead weeds stand out from the snow, spiky silhouettes of poverty weed and Russian thistle. I go outside a little after midnight to look for a moon, but there is none. I stand beside the trailer in this strange cool light, listening to the plant; the voices come seldom at night, but the roar does not abate. I look up. The light that pours on Burtco Court tonight is brought here by the plumes that race overhead; they are so low, so broad, so steady, that they reflect the plant's amber and green lights back down on the field of long, narrow boxes and snow.

Looking up, I am nearly toppled by vertigo. Suddenly the constant passing overhead of slender clouds demands to be imagined as a wake: I cannot think of it other than that the stacks and the towers are some kind of prow, cutting foam in the sky. The rush above applies torque to the ground; I can feel the earth spin. It's going the wrong way. I stumble inside and shut the door. All night my dreams are twisted.

52

It is still winter on the Rosebud. Down along the frozen creek bed near
Don Bailey's home snares are set in the tangled black brush: thin wire
loops hung from branches. Bailey and I check the snares on a black
snowmobile, hurtling across shallow snow fringed by pale golden grass.
Coyote pelts are bringing seventy-five dollars this year: The belly skin
is valuable because it resembles bobcat fur.

Bailey is a man of deceptively round edges. He is slightly plump;
though he's only thirty-seven his hair is swept over a balding spot
which it covers inadequately. He has a bounce to him; if he were a gro-
cer it would seem ingratiating, but out here it lacks the urgency to
please and it seems to be more the taut springiness of an irritable man.

Bailey runs about six hundred cows on about twenty-two thousand acres with his father, Jim; he gives me the figures without reluctance. He has either become accustomed to the demands of the press,
or perhaps the reticence of the older generation no longer matters to
him. "We've already lost our lifestyle as far as I'm concerned," he says,
as if in explanation. "But I think we're fighting for our very economic
survival here right now."

Perhaps the newly valuable coyotes, none of which have been so
absentminded today in their journeys up and down the sheltered creek
beds, to blunder into a snare, are part of his fight for economic survival. Or maybe they're just a diversion; justification for him to get out
and run the machine across the snow. He doesn't attempt to explain
his rationale for his actions as Wally McRae does. I doubt that it is
even a matter for contemplation to him. The trapping is just a part of
his continuing relationship with nature, which he calls, in about as
philosophical a reflection as he ever permits himself in our conversation, "more of a negotiation than a fight. I've never got in a battle with

Mother Nature and come out on top." There do not seem to be a lot of shades to Bailey, and he soon uses that topic to bend around again to the subject of coal.

"That same consideration holds true to some extent as far as strip mining is concerned," he says. "It's going to so radically alter the environment in many ways—I don't think they have any idea what the final outcome is going to be."

We are sitting in his dining room after our unsuccessful hunt. Bailey talks with a slightly singsong voice—it has the same kind of tension as his personality. As he talks he plays with a placemat on the other side of the table, pushing it around as if it was a piece of a jigsaw puzzle he was trying to fit into an odd-shaped hole. There is a dish of plastic fruit in the center of the table. It is beginning to snow outside the sliding glass door behind him; the red scoria hills on the other side of the creek are fading behind a slanting feathered air.

Bailey's roundish face hides a stubborn jaw. The muscles work in his cheek. He leans far back in the chair and folds his arms, as if resolution and anger tames his irritation. Connie Bailey has joined us at the table; she is a dark-haired woman with no sharpness in her manner; she watches her husband with a gentle humor, almost as if she thinks this crusade of his is only slightly more serious than hunting coyotes by snare and snowmobile.

"I'm sure that history is going to prove that we are right," Bailey says, clapping his pointed boots together under the table for emphasis. "But there's not going to be a great deal of satisfaction in telling anybody I told 'em so when there's a town of twenty-five thousand people over at Colstrip and smokestacks and all that goes along with industrialization; when plants and stuff are dead around here because of the pollution and little shack camps cropping up in every coulee." The restraint of anger seems to be a common trait among these ranchers. But it seems that it is more of a struggle for him than for Wally McRae.

Have you ever thought of using guns to defend your land? I ask.

"Damn right," Bailey says. "I mean it. I've already told the deputy sheriffs here, when you guys come out here with a court order telling me to step aside and let the core-drilling trucks go in, you'd better be goddamn sure you want to deliver that court order."

Connie giggles, as if slightly afraid of the threat, then tries to limit it with explanation. "He's—" But Bailey continues without a glance at her.

"I have said that if I don't feel we're getting a fair shake we're going to fight. I mean that literally." His boots whack together hard. "I don't mean just we're going to be obstinate. I mean we're going to fight."

Bailey's land sweeps up into the Sarpy Mountains. We drive up in between fingers of red-topped hills in his four-wheel-drive pickup, and Bailey fights the drifts as if they conspired to slow him down. A small creek flows down the floor of the valley, but even the darkness of its open water against the snow seems pale compared to a slice of coal lying like a cave along one of the banks. It grins, toothless, at us as we drive past.

Within close view of the dark, white-stippled forest that tops the mountains we are finally stopped by a drift that blocks the road, polished by the wind, solid as pale granite. Bailey does not even attempt it. He stops the pickup and points. Over there, he says, there is another mine just five miles away.

The land around us is absolutely still, crystallized by winter. We have left the last cattle far behind, nosing around in new-fed hay for the last chunks of cow-cake. The unbroken snow stretches into the hills in all directions. Except for the pickup, this could be the winter of 1876, when the Cheyennes and the Sioux, perhaps guessing that they had fought their last great victory, wandered with little purpose across their land, like old men visiting rooms where there had been love. It is almost impossible to believe that we are hidden from the clank and turmoil of a strip mine by just a mountain and not a century. Bailey and I sit in the cab in silence for a few minutes, until I can almost imagine that I feel a faint rumble through the earth. Unseen, the mine chews vastly into the other side, and the view of the eternal mountains goes stale.

Bailey spins the tires and slews the pickup around, and we smoke back out over the hills, our ride made more interesting by those ever-present chunks of frozen manure. And again I feel Bailey's anger at the encroachment of the miners, moving behind the mountains to consolidate territory and peel it open. But I am finding that it is a different kind of anger than McRae's. And the difference goes back to their attitudes toward the land. McRae's is almost mystical, a melting of heritage, blood, history, freedom, and livelihood into a core of feeling that is the only thing he has trouble articulating. Bailey's is utilitarian. The land has his respect, but perhaps not his love. As we jolt down the road and plunge through the drifts we hacked open on the way up, I ask Bailey why he enjoys this way of life.

"Oh, I think it's—" He pauses, thinking it out, then falls back on a platitude. "It's a continual challenge to adjust to the whims of Mother Nature, I think, you know. Everything you do with your cattle, you of course do it looking at it from an economic standpoint, and the one thing we can't control in our environment is the economics."

We return from our journey into the Sarpys to feed a small herd of weaned bull calves. These are precious creatures, dark, husky Herefords, bearing the blood of countless future heifers and steer calves. These are the class of Bailey's herd. This side of the operation is where he can focus all his education and knowledge in genetics, breeding, feed, and care, to produce the quintessential bull: long, tall, well balanced, incredibly broad, with testes the size of bowling balls and a languid nature that packs all energy into reproduction. Out on the range with his cows and calves Bailey is little more than a glorified herdsman, like all ranchers; but with his bulls he's a scientist, weighing grams of ancestry and innovation against size at birth and growth performance, and the quality of meat. Most ranchers seem to need this kind of sharp edge to keep themselves keen. Perhaps it is symbolic of the difference between the two men that Bailey's edge is raising bulls, and McRae's is breeding quarterhorses.

A few days ago I asked Don Bailey and Wally McRae to describe for me what they would be doing on their ranches now if the coal controversy had not bloomed as a central focus of their lives. It was at McRae's home, and both men were tired, but that didn't keep the two of them from developing an argument out of what I at first thought was an insignificant point.

They had started in tune, like a well-practiced duet. They tried to explain the worries and concerns they had before coal came along to diminish everything else. They talked about the dipping vats they built on each other's ranches; they described some of the routine of going from branding to haying to selling to feeding to calving year after year. Right now, Bailey said, he'd probably be clearing land of brush if he wasn't spending so much time on the coal circuit. McRae said most likely he'd be trying to figure out an irrigation system for some land he owns on the Tongue River down toward Ashland.

Then they were talking about what makes an economic unit—a

piece of land on which a man can make a living following a cow around. And Bailey kicked it off.

"You know, Art told me the other night, and he was right. He said there's not a viable operation in the country that doesn't have outside income, and he's probably right. Over a period of forty years every one of these outfits has had some source of revenue other than the cattle that we're raising on the grass. Might have been an oil lease or a timber sale or—"

McRae interrupted. Those thin eyes flickered.

"Yeah," he said, a little warning in his voice. "But it all went back—the land. All of these things go back to the land."

Bailey plunged on in, his thick forearms resting on his knees.

"But we are talking strictly about the cow business. Coal goes back to the land, too."

McRae did not agree. His voice, well trained to the rodeo arena, rang with disgust, much of which, I suspected, was synthetic. "Well, you'll have a hell of a time ever staying in one of these outfits if you don't have the cows, though. The cows are what keep you in the damn business."

Bailey was dogged. "Well, how are you to know?"

"Oh, hell. You can't operate that thing on what some oil broker's going to give you." McRae may have been sensitive about the oil exploration lease he had signed a year before. It had helped him get through a year when cattle prices were once more below the 1951 level, and he may also have hoped that the oil prospecting might help confuse the coal issue.

"Well," Bailey continued, gnawing the argument down to its slender marrow. "If you devoted the attention to exploiting or developing your other possibilities you might be able to. Cows are a liability about eight years out of ten."

"I don't know what the hell I'd sell around here if it wasn't cows. Homesites?"

Bailey, with no liberating streak of humor, answered:

"Deer."

When McRae didn't laugh I realized that this argument must be hitting something fundamental. The idea of the lean, hard Wally McRae herding, branding, and feeding a skittish herd of does satisfied my yearning for farce, and I had to reach quickly for one of Ruth's cookies to remain sober. But McRae never broke.

"Deer," he said. "How many deer have I sold this fall?"

Unmoved by sarcasm, Bailey went on. He was just warming up.

"If you wanted to devote the attention to promote the thing properly there's any number of things that you could develop to make money, and they might make you a hell of a lot more money than cows. You just haven't devoted that kind of attention to it."

McRae let the silence grow after that one, so I filled it with a question about history, and the conversation cooled. But later I wondered why it had been so suddenly hot. Perhaps, I thought, it was because Bailey had attacked the purity of a way of life that to McRae, the romantic, was precious beyond reason, but that to Bailey was only a challenging way to make a living.

Bailey does not just feed his bull calves hay. He gives them corn and grain and mineral supplement.

Wearing a short-brimmed cap with ear flaps in the cold air of a shed, he works amiably at the job of preparing the mixture, shoveling grain and corn and mineral mix into a machine called a Gehl Mix-All where the dry amalgam whirls behind small Plexiglas windows. He feeds the corn in through a chopper that whines and hacks, occasionally throwing out kernels as if it were spitting seeds. The grain is lifted in through a tube by an auger. The corn is deep gold; the grain pale; the smell is of a dusty ripeness. A fine powder, ground by the churning of the mixture against itself, rises in the air and drifts into the clouded light by the door of the shed, and I watch Bailey through a silver haze. Out here his tension seems to be gone; the routine has drained it away. He turns off the machine and reaches for a handful of the feed. In the sudden silence he shows it to me: a rich, dry mound of kernels and cob and husk and grain.

"That's real calf feed," he says quietly, and carefully throws the handful back into the bin.

Driving out to the calves, Bailey remains relaxed, almost philosophical.

"I used to think, before all this came about, how fortunate we were," he says. "That probably we had the quality of life that, maybe, our forefathers had in the backs of their minds when they came to America, and that probably I came as close to realizing those dreams as anybody. When my dad was a young fellow, growing up in this country, he had the ecological goodness that we have today, but he didn't have the standard of living that we have."

But after the calves are fed, lined up as symmetrically as a row of paper dolls at a long, sectioned trough, the subject returns to coal, and Bailey's ease of manner deserts him. And I begin to understand the restlessness that nags him. Unlike McRae, who has no plans but to stay put and fight, Bailey has already begun turning his options over in his mind, and sometimes the possibility of being free from this turmoil and richer than he ever imagined must feed longing.

Freedom! Who else in this nation must put up with this enduring pressure: the visits from the coal buyers, the core drillers, the journalists, the fear? Who else has to live each day, work each day, caught by this uncertainty that runs, like a wick, like a fuse, through all life, and into the earth itself? Trapped by a quirk of geology the way the rocks themselves trapped the coal. Freedom! To worry about bulls and hay and winter and market prices, just these alone, and against a background of serenity. And what peace! With work and duty satisfied, to sit in a quiet house, with no meetings to attend, or calls to allies to make, or strategy to plan, or lobbying to do, and to know you could grow old there, nestled in routine. Freedom!

He has a specific dream, a safety-valve plan, which probably irritates his restlessness as much as it relieves his pressure.

"I think," he says, "I'd go and find a nice quiet coulee somewhere and run two, three hundred of these registered Hereford cows, and go there and have enough extra income for whatever we get out of the thing. I told my cousin Jack that we'd buy a big fishing boat and put it in the harbor at Sitka and go up there every spring and fish all summer and come back down here in the winter time."

Bailey is quiet for some time. It has begun to snow again, very gently, each flake independent, fading the Sarpy foothills into a charcoal sketch.

"I don't know when a person reaches a point where the economics of the thing overrule your convictions. I'm sure that everybody has that point, though."

There is a story, famous in eastern Montana, about a rancher in the Bull Mountains, also threatened by coal development, who told Consolidation Coal Company that no matter how much money he was offered for his ranch the company would always be $4.60 short of the price. Bailey, pragmatic man, may tell the story as useful rhetoric, but perhaps he doesn't believe it. When I ask him if he means that everybody has his price, he says:

"I think so. I do. I think that when everything is weighed I think there eventually is a price." He pauses again, then begins to qualify himself with conditions—the pressure, the economic situation, the closing of alternatives. It seems that he is now hardening himself up after a moment of weakness, the excursion into his longing for freedom. He is putty, he is clay, he is stone. The pickup bangs on the rough road. His jaw muscles work.

"But if I have to leave here," he says, cold as marble, "my financial future is going to be secure. There isn't going to be any question about it. I'll be a reasonable man as long as I think that there's something in it for me, but when our back's to the wall they're not going to go in the country up there until the deal suits me." He grapples with the truck through another drift. "If they want a range war they can have a range war."

53

A few days later it is so warm there is no steam. Spring is coming to Rosebud County. Most of the snow has gone, leaving the earth looking pale but clean. Even the pink scoria of Burtco Court seems fresh on this warm morning. When it is this warm and dry the steam from the stacks does not condense into vapor clouds as it blasts upward; it remains invisible: All that can be seen is a thin plume of smoke, like a faint exhalation of dust, rising from the stack. Unlike the steam that gushes out in cold weather, however, this plume does not dissipate. Today the air is so still that it accumulates down to the southeast, and hangs against the horizon in a very light gauze of red haze.

It is another day of hope for the ranchers of the Rosebud. Today, it has been rumored, the presidents of the five companies involved in Three and Four are meeting in Portland, Oregon. It has been said that this meeting will determine the future; that the delay in the start of construction has been so protracted and the date on which it may begin is so uncertain in the face of the Cheyenne request for clean air that some of the companies are ready to pull out and make other plans. As usual there has been little talk of this hope among the ranchers, but I can feel it in the air this time. Perhaps they instinctively believe that this kind of conclusion—a gentle dissolution of the consortium—is more likely than any kind of pronounced legal victory.

Knowing this, it seems to me that the day, the weather, even the power plant, seem poised, almost as if all held their breath, waiting. After the past cold days, this warmth is soft; there are a few little lazy clouds adrift in the sky; even the black dust that always rises from the coal stacker beside the plant and usually seems to writhe into the air, today sifts quietly to earth. Even the town seems subdued; I walk over to the Colstrip Community Center, a gym and a swimming-pool com-

239

plex, and though there is a distant hollow whacking of someone playing racketball in the basement, no one is playing basketball in the gym.
It is as if spring had come early and caught the sap of the tree sleeping
deep beneath the bark, and we all stretch and blink and sigh with
pleasure to slowly wake. I write notes on the day, convinced that the
Colstrip decision will be made today and I will want to remember this
atmosphere.

I drive down to Wally McRae's house engulfed in this feeling of
history pending, and find him working with his horses. He works with
such a tight energy that he seems angry, running and flicking a short
whip. But I have seen this mood before; it is his style in the corral:
crisp, authoritative, intense.

It is calving time on the Rocker Six. Most of the calves are born in
the pasture, produced by their stoic mothers with little fuss; but
McRae brings many of his heifers, who have never borne calves before,
into a corral near the house. There he can help them if they need it.

McRae invites me to dinner, and by the time it is over the night,
still coming early, is full. There is no moon, and the sky is slowly cooling into a pure black. McRae takes a stainless-steel bucket and invites
me to go out to the corral to watch him milk and see if any calves are
coming. As we pass through the gate he flips a switch on a pole and
lights the corral. Two or three heifers are in there now, and one is lying
near the center. She looks over her shoulder at us as we pass. "Looks
like she's near," McRae says.

The milk cow is in a stall. McRae sits down beside her and begins
to squirt the milk into the bucket. I stand in the shadows absorbing the
slowly changing sound and the rhythm of milk on the tin, milk on
milk, milk on foam. I know that by now the decision has been made in
Portland. There is a sense of remoteness here, of living on the prairie in
1884, waiting for the election returns. How soon will we know?

McRae's head is pressed against the flank of the cow. He has told
me before that this time with the cow and the milk is his time of medi-
tation, so close to the old world of the ranch: the soft warmth and
breathing of the cow, rich smell of her manure, the hard warm teat,
with clumps of milk passing through under his hand. Here he has
watched a cat writhe and purr and go through a whole routine of worship in its desire for a lick at the warm milk, and he has thought:
"That's just like religion. The cat thinks the reason he gets fed is because he does those right things." There are no cats now—for all their
faith, the hired man's dog has killed them.

McRae carries the bucket of milk back through the corral. The pregnant heifer glances at us again, gets up, and jogs heavily over to the other side of the corral; in the light I can see she is bleeding. McRae opens the gate, waits until I am through, and turns off the light. The dark is so suddenly absolute I stumble in the soft, lumpy earth. I think of the light and of the little electric heaters most ranchers have to keep water liquid for the cattle through the winter.

At least the coming of electricity helped the rancher, I ask. Didn't it?

Almost automatically McRae replies: "Yeah."

We walk in silence. He knows the way back to the house by heart. My eyes now blinded to the immediate front by the single distant yard light, I follow carefully behind him, imagining hay rakes in the grass.

"Maybe," he says. "No. Maybe not. Maybe it was better then."

I follow him back to the house. He sits in his chair at the end of the table, lights a cigarette, and again we talk. He seems subdued. At last I ask him about this victory that seems so close. He sits back in his chair, his eyes nearly closed. The house seems empty. I can hear it ticking as the warmth of the early spring day slowly seeps back into the sky.

"It's too bad that somebody has to win and somebody has to lose," he says. "That's the kind of thing that Martin and I would like to avoid. Because maybe they could have built Three and Four in a place where everybody could have won—Butte . . . Seattle . . . Great Falls . . . even Billings."

For McRae this is a rare mood. Before the triumph is even official—even real—he is reaching in his mind for the proper attitude of conciliation with which to try to mend the splintered morale of his county. What does he see as he looks down into a future without Three and Four, and, more importantly, without the hopes and fears of its possibility dividing the county across relationships that could otherwise grow? Perhaps to fully mend that chasm of bitterness there are compromises worth making. I ask him: The cancellation of Three and Four still wouldn't solve your problems with coal mining, would it?

There is a long silence. Then he says:

"I've done some numbers on Martin. And Martin's done some numbers on me. And maybe all of us do have an obligation to the— quote—energy crisis. But it's a limited obligation. It's pretty damn hard to mine coal when there isn't any coal. And there's coal here. If the companies would try to mine coal as responsibly as they would lead the public to believe, there *might* be an accommodation between the cowboys and the coalboys." He pauses again. "If they'd just stop building these things here, I think Martin and I have sown the seeds of a community."

54

The town of Billings smokes pungently up into the big sky. It, too, has felt the impact of coal development; it, too, is booming. A new shopping center completed a few months ago is the most obvious symptom of this boom and its cause: The center, Rimrock Mall, will use a healthy percentage of the entire output of Montana Power's Billings power plant.

Sheila McRae is living in Billings now, going to school at Eastern Montana College. I have come up here from Colstrip to interview her and two others. There has been no announcement from Portland, one way or the other; the companies are holding their fire, or their flag of truce.

I find Sheila McRae in a small apartment a few blocks downhill from the place where the college nestles under the wall of sandstone that rims the valley. She sits in a small room in front of a poster of an Indian smoking a joint. The words on the poster say: "If Custer had only known." She is tall; her face is almost a perfect oval; her hair is tied in twin braids by multicolored elastic bands.

My questions are longer than her replies.

Are you studying anything in particular?

"No."

Do you want to get a degree?

"No."

Do you ever want to go back and live at home?

"No."

What are your long-range plans?

"I don't really have any."

What do you feel about Colstrip Units Three and Four now?

"I don't know."

243

I am stifled, cut off in my inquisition. This is not what I expected. I thought I would find vigor and flashing words; opinion, fire. In the vacuum of their absence my questions stall.

There is silence. She watches me placidly, a slight smile scarcely creasing the fine, pale complexion of her face. Her eyes are awake but unstirred. Her hands are folded. She is still. She waits, courteous, wary.

I am bewildered. It has been a long time since I encountered such a passive quietness in a person nineteen years old. It is like hitchhiking to a rock concert and finding instead a single cellist in an empty room, playing sweetly for his own ears. But I begin again, hoping there is more in her than silence can hide forever.

I am curious, I tell her, about the effect of the power plant on people growing up with it. Everyone I have talked to so far has come to the conflict with prejudice. I wonder how the unformed eye has seen it, being formed by its power. Sheila had just turned twelve in July 1968 when the first Montana Power coal was shipped to Billings. She was seventeen when she spoke in Forsyth. It was part of her life for those vital years—the construction, the coal trains, the young men who come and go, the clash between rancher and worker that somehow trapped the children, crouching low, under the curved red arc of anger.

She seems to know what I'm trying to learn, so I ask her again what she noticed when development began. Drugs, she says.

"Even if One and Two hadn't come there could have been drugs," she says. "But I think there are a lot more there now than there would have been."

How did you first see it, I ask.

"Kids coming to school loaded and talking about it, and seeing people at parties."

Did you get involved?

There is hardly a hesitation, but her eyebrows defy me to be surprised.

"Yeah."

Just grass?

"Yeah. I first did it at the end of my sophomore year. 1972."

Was grass the dominant thing?

"Yup. But . . . Well, there's a lot of hard stuff going around there, too. I didn't know that then but there is now, I know."

How do you know?

"Friends of mine that do it."

Do you agree with them doing it?

"Yeah, I think so . . . Maybe . . . I mean . . . I think . . . I was talked into trying grass and I haven't tried anything else but I'm not saying I couldn't be talked into it. And they obviously were and I'm not saying I couldn't be."

I am stopped in my questions again, wondering what conflict between her reality and her image as she thought her parents saw it made her speak that night in Forsyth and sound like a child dismayed and above the degradation she saw around her. Or was she angered or regretful at being exposed to forces that she could be talked into? Was she afraid of the flood of marijuana and cocaine and heroin that came to Colstrip with the construction boom, wishing, while she sampled the fringes, that it had never come to offer her its fluid door into a world of shadows? No wonder there is hesitation in her words and a little fear at my questions. She doesn't know where she stands.

I back away from her conflict. I ask how the school changed. She is vague. "Seemed like when I was a freshman and sophomore, there were a lot of dances, but toward junior and senior they weren't having hardly any. Every year I looked forward to the school picnic at the end of the year. We always had it. Everybody came and after we'd turn in our books then there was a big movie, a family movie, and after that we'd get out, get our report cards, and there'd be a big potluck; all the families would bring stuff in the gym. They'd all be lined up, all the parents and everybody'd be there. And that was from the first grade till my sophomore or junior year.

"And I don't think they have it anymore. Nobody around to bother with it anymore."

Sheila does not like to be asked where she is going, what she is going to do with her life. She shrugs it off, and prefers to talk about someone else.

"I think of one guy in particular who came," she says. "Who really hated a couple of people because they said that he came in, you know, and he thought he knew everything. And he was really upset that people would feel that way because he was just new. He said he hated moving—he hated the thought of moving to Colstrip, because he'd heard it was a terrible place. He liked it all right, I guess, once he got here. . . ."

And in a moment or two more I understand. Out of sympathy grows friendship, and out of friendship, more. Sheila wants to talk about someone else because she can't help thinking about him. When I ask her if anyone's relationship to his or her parents was changed by the power plant she says: "Yes. Mine. Because of the guy I started going with. His father works at the steam plant."

No wonder she's ambivalent about her opinions, I think. Her heart belongs to both sides.

The day after the Forsyth meeting Mike and a friend spent all their time arguing with her. "They didn't change my mind," she says now. "But I started being a little more broad-minded than I was before." So now she tries to avoid taking any position.

"When I talk to Dad and the ranchers I would say no, I don't want the steam plant because it's going to . . . that land is partially mine, too. And when I talk to other people who are making good arguments for it too and I can see their points and I can see Dad's, too."

Besides, she is reluctant to take a position on the future of Colstrip, she says, because she hasn't lived there for a while. Since she graduated in 1975 she has lived in Billings, Portland, Bozeman, and Billings again, almost as mobile as a boomer. And Mike married another woman.

"I don't think Colstrip's going to affect me that much from now on," she says. "It's going to affect Dad and the ranch and the boys probably because they're going to be there, but it's not going to affect me that much."

You don't plan to go back?

"No."

55

Colstrip One and Two consumed Robert Olmstead and ejected him. Olmstead didn't survive the start-up. Before the plants were fully on-stream he had been sent back to Billings by the company, taking his family with him. "At the time so many people came to say that it was a natural thing, that every time they build a steam plant they can go through three or four superintendents," Lois Olmstead remembered. "And then it comes back to his personal faith in the Lord that nothing happens in our day that he doesn't allow and if it meant going back to Billings, then that was fine."

We talk in her home in Billings. Robert wasn't fired; he was just demoted in megawatts, from Colstrip's 700 to Billings' 180. We eat sugared doughnuts in a green room with green carpets and green drapes. There are open Bibles on the end tables, a gleam of gold-edged paper under lamps hung with lobes of crystal beside a fireplace of dark wood and brick. Soft sunlight in the drapes and the ebullience in the woman brighten the room.

Lois Olmstead bubbles with unforced joy. She is happy to talk about her experience in Colstrip. She's an opposite to Sheila McRae: I have to fight to get my questions in, and in the course of an hour I probably am only allowed five. She fills the rest of the time to the brim. The most important thing to her was that rancher and plant worker could get together in peace and friendship behind the clatter of their public dispute. It seemed to her that in light of that accomplishment the issues weren't all that important.

"We decided to have a Sarah Coventry [a jewelry brand] party," she says. "This is really"—she laughs—"but this is the way it was, I have to tell you. We decided to have a Sarah Coventry party and Iso-bel [a long-time Colstrip resident] invited all her rancher friends and I

247

invited all my friends from Colstrip who were power company people and construction people and so I had this party in my home and then all these people came together and I can't say—it wasn't like a normal everyday party and I really wasn't apprehensive—I guess I was excited about the ranch ladies meeting the city ladies, if you put it that way. It was in my home and I suppose there were twenty-five ladies—and we just had a really good time visiting—I can't remember you know, all the ranch ladies sitting on one side and all the other people on the other; there was none of that. It was women together and they were all gabbing. I remember that Ruth McRae came to the party and I was really excited about meeting her because my family was also rodeo background and all of us were involved in rodeo and I knew Wally from that—you know, not as a person—but from announcing at rodeos. Oh, yes, and because he had been on TV and we took the [Billings] *Gazette* and read it every day and I read everything I could on the development and kept most of the clippings and that kind of thing.

"And one day after this Sarah Coventry party we had to deliver all this jewelry to everybody so Isobel said let's pick a day and we'll go out because she wanted me to meet all of her friends—so we first went to McRae's and I was thrilled to death to get to be in her home and you know and to develop a friendship. She was fixing a lunch for a cattle drive so we didn't meet Wally—but we visited with Ruth and I saw so much likeness between her and I because she would see the people in the situation, you know, which is opposite to—in contrast to— the big issues, there's people involved. . . . Being from a ranch background I was so interested in how they felt and yet I wanted them to see we weren't, you know, villains in to snatch up their land, that we were *people*."

The Olmstead's home is just down the street from Rimrock Mall, the vast consumer of electricity. Lois tells me how irritated her husband was with that level of consumption, how one year he wouldn't let her put up the Christmas lights on their house because of his concern for conserving electricity.

"We hear about the problems firsthand because Montana Power Company either sells or gets power back and forth, and Robert will come home and say today so and so was short of power and had to buy so much, because they couldn't handle the load or Butte calls and they have to crank up the steam plant because you know everybody needs it and man we've got to keep this thing running when it's really cold and the pipes are freezing up; he'll maybe be down there all night making sure that it's running, so that's real to me. We've been turning our heat down now because of what we hear, but there's so many things that are faults and he knows about it—like shutting lights off is a hole in the wall compared to how I use my dryer or my washer, and he gets so upset over these people you know who turn lights off because they're going to save electricity when that isn't what does it, and he marches around the house every now and then when he thinks about Rimrock Mall and the electricity it uses or goes into stores that are so hot when everybody's got their coats on. The way we feel most definitely is if we don't save and if we don't do our share then our kids will suffer the consequences."

Lois Olmstead didn't speak in public again about the plant after she spoke at Forsyth. But she went to another hearing, this time in Billings, and there, at the end of it, she found again the core of experience that gave her beliefs substance—human decency overcoming controversy.

"We went to the Billings hearing and I thought if I get a chance to say anything I'll be glad to say something because I want the people to know that these ranchers are complaining about things that ranchers are having trouble with all over. My dad was having the same complaints at Livingston which was what, two hundred miles from Colstrip—increased traffic and hunting and that kind of thing—and I thought if I get the chance to say that, I would love to have that go down in the paper. Well we came to the hearing here and I raised my hand for about an hour and a half and they never did call on me. Then Nick [Golder] stopped and visited with us after, and we visited back and forth with some of the ranchers and it didn't get over until like 1:30 A.M. We went out with Don Bailey and Wally and Ruth were standing in the hall talking and they said, 'Well, are you going to have some breakfast?' And we said, 'Well, we'd better as long as this thing's been,' so we all went to Sambo's and had breakfast together—I thought, *drats*, I wish the public knew that we were friends."

56

Clifford Powell smiles like a man building a stone fence. Although the smile is not unfriendly, it's a barrier, and it's hard to see through. His bright, even teeth come up and *slam,* there is a flash of the mortar of good humor, and then it's over.

Powell has been a pipefitter for about twenty-two years, ever since the man who was paying him forty dollars a week for his work in a bakery offered him a generous raise of a nickel an hour and Powell decided to leave to become an apprentice. He went down to work at Colstrip from his home in Billings because he had never worked on a job like it and he wanted the experience. He was a general foreman in heavy rigging and critical pipe before he came home to take the job as business agent for the plumbers and pipefitters union. I interview him in his Billings office, where he sits in shirtsleeves, a small notebook in his left shirt pocket. He is uncomfortable with the interview. He pushes his desk drawer in and out as we talk.

We discuss boomers. He dislikes the word. All those people are, he says, are men who need work.

"Criminey, I've had to do it—I've worked on the missiles in South Dakota and the atomic facility at Hanford, Washington—just about everybody in the trade has had to do it. Well, criminey, that's not an enjoyable life for a lot of people, but they have to do it. To me they live a hard life. They've been called tramps, transients, everything else, but the world better be damn glad these people are willing to do this kind of work."

Powell is playing with a steel ruler, bending it back and forth in his thick hands.

"It seems to me like especially your farm communities have got an isolationist point of view as far as what our duties are to the people around us. One of the basic arguments is that Montana doesn't need

250

the power so why should we produce more than what we need here, and I think that's narrow-minded."

But Powell has other concerns right now than making sure the country doesn't run out of power. He is faced with a problem fundamental to any union that has benefited from a large construction project: the terrible fact that the project was temporary. With One and Two finished, he has about a hundred men out of work, and Units Three and Four would yank them out of unemployment like a wartime draft. He's in the old squeeze: Once you've had a big project, it's so hard to get along without another.

Just before Powell went down to Colstrip in 1973 the pipefitters were in good shape. "We were busy and we weren't in any bind— everybody was working."

During the construction of One and Two the union added a number of new plumbers and pipefitters to its membership, many of whom came to the area from other states to work on the plant. In 1973 the union had about 200 members, and now, Powell says, it has 307.

"We probably had about thirty-five or forty members of our local on One and Two," Powell says. "We had 150 or so permit people, and then there were probably 150 people that traveled from various parts of the U.S. We had guys from New York and Florida and California and Canada. We had a lot of Canadians down."

But when the construction ended and the visitors went home, the union was still left with that extra hundred.

"Really," Powell says, "those people we took in because of the project is about the amount that we've got unemployed. In fact, this is the worst year that we've had in this city."

We talk about unions, and their changing role in American society. "The role of the union today is to improve the livelihood of people making minimum wage and below, and to make sure people make an honest living and get fair treatment from their employers. Unions have a bad name; people are quick to point out they're corrupt, but, criminey, when you look at the corruption of big business. They buy everything."

So I ask him if it doesn't feel uncomfortable getting up in front of a group and publicly aligning oneself with a major corporation like Montana Power Company. The company recently had a squabble with the head of the Montana AFL–CIO, who accused the company of threatening to use nonunion labor on Three and Four.

"Yeah," Powell says. "Montana Power is not pro-union by a long shot. But we want the work, we can't deny that. We want the work like the farmer wants high beef prices. Montana Power's spent their share of money battling us. And we've spent a lot of time and effort trying to help them on this project. I'd hope their attitude would improve."

Powell smiles, a short, flat gleam of teeth. The interview is over. He must get back to work. He gets up and shakes my hand and walks me to the door. His big eyes glitter; he gives me another smile. His voice grows quiet.

"Now, we're just talking," he says. "But you know the way to weaken our country is through our economy, and the way to weaken our economy is through our energy, so don't you think there might be a conspiracy somewhere? You can look all over the country. They're holding things up all over with small groups. They don't want coal-fired plants, water projects, oil-fired plants, atomic plants, refineries, pipelines, offshore drilling, strip mining—*any* kind of mining. I'm not saying these guys here, Northern Plains and all, are *communists,* but—they've got to get their money from someplace."

Powell smiles and closes the door.

57

Myron Brien goes back to the Black Hills to play; Bill Parker went there to get married, because there was the mystical heart of his tribe. His second marriage was in 1974, the same year as the Forsyth meeting, a year of emotional exaltation for Bill Parker. It is possible that nothing since has come close.

I find Parker in a small trailer in Helena. I have come here from Billings; I have not been back to Colstrip. I have read nothing about events in Portland at that distant meeting that was to tell the future. It was days ago now, and there has been no word, no sign, no flutter of withdrawal from the companies. Limbo extends, but surely it is like a single piece of elastic and will not expand forever.

Helena is the capital of Montana. It lies huddled against the cold side of a mountain like a dog nestled into the snow for warmth. The capitol building looks north, at the source of the wind. Like Butte, Helena is a town built on mining, a city on a hillside, rich in old brick buildings; unlike Butte it has survived the mining with charm. Renewal of the downtown area has preserved some of the atmosphere of the mining city while removing some of the grime, leaving a romantic memory of times that were never really very good. Unfortunately, the charm does not reach out quite as far as the little trailer court where Parker lives with his wife, Sue, and their two children.

Parker is a short man, slightly stout. He is forty years old. He has a high forehead across which squads of furrows march when he frowns. The Indian in him is subtle: In his face it is texture, not color, a finely coarse roundness of feature, as if he had been carved by blowing sand. In his voice it shows slightly in tone and pace. He is caught, too, by a sense of language, which shows most in his written testimony; a sense that is reminiscent of the eloquence of the Indians of the past and is less evident among tribal leaders today. I mention that he seemed to be one of the more eloquent speakers at the Forsyth meeting, and he grimaces. The furrows march. "A gift of God," he says, but without any apparent pride. "An excruciating one."

In the trailer a heater moans sporadically. Sue listens to the interview with a faint smile. When I ask her to describe her husband, she says: "A dynamic person. Real creative. He has the ability to reach out in unexpected ways."

"Irascible," Parker says.

"Oh, that, too," says his wife.

We talk about his past, about his life in Los Angeles, where he was poisoned by sulfuric acid; the memory makes the thought of acid rains precipitated out of stack effluent particularly poignant to him. When he came back to Montana in 1964, and he began to study Indian history, he recalls, "I went through a Red Power phase. Most young Indians do. I knew it was bad, but I didn't think it was this bad."

It was then that he began to get the feel for historical rhythm that still drives him.

"One hundred years ago we were central to the economy of the United States," he says, leaning forward, elbows on his knees. "The country was on the gold standard and we had the gold. And it just so happens that one hundred years thereafter we're on the energy standard, and we begin to see that in every regard the Northern Cheyennes are the hub of the wheel."

He is vehement. The furrows in his brow deepen; the language pours out of him like sweat.

"For years the Northern Cheyennes lived in a suppressed state in the Powder River Basin. Then all of a sudden you find yourself confronted by Consolidation Coal, Peabody Coal, American Mining and Exploration Company. You find yourself confronted by people like that who want your land. You wonder where there is any justice!"

Parker leans back on the sofa and folds his arms. "One hundred years ago we were finally given a small piece of our land. But one hundred years ago a man named Edison was tinkering around with light so that one hundred years later we could be assaulted again."

Parker is in Helena because his wife works here, and he has been working here as a tribal lobbyist, and there is no work for him on the reservation. He is now unemployed. "I'm just coming off the bottom again," he says quietly, talking with his eyes closed. He doesn't describe bottom. "There's been a number of setbacks that take your confidence away. But things are on the upturn."

Still, he is not sure which way things will go. Heritage today is a popular possession, and to have colorful ethnic roots is more romantic now than to be descended from Mayflower stock—today strange young white men, desperate to adopt a culture, come to the reservation dressed in home-tanned leather and smuggled eagle feathers to live in canvas tipis and try to get to see the Sacred Hat. But to be Indian in possession of valuable resources is not necessarily a comfortable thing. Whites cannot help wondering: Is clean air important to the Northern Cheyennes for itself or for its political value as both a unifying issue and a weapon against oppression by the United States? Bill Parker has no profound answers. I suspect that to him being a part of the tribe is enough, being a part of a small indefatigable nation that made history a century ago and, he is sure, is soon to make history again. His deepest fear must be that in the end the Northern Cheyennes will not fully accept him as their own.

Parker sits on the couch, and plucks a little piece of lint off his trouser leg. "We're not permanent here," he says, looking around the trailer. "I keep myself loose for my tribe. To go whenever they call."

58

In anticipation of what one company spokesman foresaw as a "major power short-age" in the area, five Pacific Northwest Utility companies have upped the ante in what may be the largest gamble in the history of the West.

On Friday the companies met in Spokane and decided to proceed with con-struction of Colstrip Units Three and Four, in spite of several court appeals and government inquiries into the project.

In the Montana Power operations building adjoining Colstrip Units One and Two, a hastily convened crowd of plant supervisors got the word from [a] Montana Power Company senior vice-president . . . on Monday afternoon. At about the same time a telephone call from Western Energy President Paul Schmechel alerted town management.

By 3 o'clock that afternoon, concluding what appeared to be a well-orches-trated procedure, a battery of earth movers parked just east of One and Two began to scrape away dirt which lies above a coal seam which will eventually be mined to make way for the additional power generating units.

One observer said it was as if the construction people were anticipating the command of "gentlemen, start your engines."

Tuesday Wally McRae said he saw no reason for the companies to proceed with construction at this time. . . . McRae said there was a double standard where ranchers and other small businesses must wait for court action before undertaking major projects, but that power companies are apparently immune from the threat of litigation. . . .

On Wednesday groups who have opposed the Colstrip project filed for a court order to halt any construction.

—*Forsyth Independent*
June 9, 1977

All the way down from Forsyth, I do not see the power plant. I have been gone too long, I hardly remember where to look. Five minutes out of Colstrip I still see no sign of it over the brown and red ridges to the south, although the transmission lines have now appeared, marching across the horizon. "Steel soldiers of progress." Paul Schmechel's line haunts me. But the plant is invisible. No cloud, no dust, no smoke. It is almost as if it has been forcibly removed from the landscape.

Then, just as I see the top of the stacks and the little daytime sparkle of the strobe lights, I remember: Of course, it's summer. It's warm and the steam is invisible in the middle of the day. So I come around a corner and there it is, green and pink and silver, still roaring.

It is August. I have been gone for over three months, working on other things. Now, as I drive through town and down past the green of the city park, across the tracks, under the power lines, and up the stony road to the place where my trailer stands, a dirty white oblong against the red rock of the slope, I look over next to One and Two, curious to see what rises there on the place that was empty so long. Nothing. The space has been cleared of the debris that littered it during the construction of One and Two, and has the freshly peeled appearance of earth that has been graded, but there are no caissons, no columns, no steel. The One to Four complex remains an unfinished idea. Incomplete, indeed, but it is a far more durable idea than I had thought on that warm spring day I talked to Wally McRae and I thought that it was dying.

I unpack in familiar noise. Roar, thunder, hiss, scream, AXZRND BLAVRZ, LINE TWO. I feel like a boomer, part of the flotsam drawn to this machine. Carrying my suitcase from the car I notice that the double-wide trailer that occupied the space just across from me up the hill has disappeared, and so has an elaborate streamlined travel trailer that had seemed permanently installed down toward the plant from me, with fences and sheds and even a small garden. All that is left is that rectangular piece of soil, like the mound over a trailer's grave. So I know without asking anyone that Three and Four are not really under construction yet, or Burtco Court would be full.

When I saw the news stories on the beginning of construction I had called a few people. "Hell," Don Bailey said. "They had a permit to mine the coal there anyway." Wally McRae turned the event around so that instead of being a symbol of the companies' power and momentum it appeared a bluster of insecurity, like the last desperate act of defiance by a besieged city.

"They're whistling in the dark," he said. "I think they're having trouble holding their own troops." But the consortium remains intact, and the hope that seemed so strong last spring that the project would collapse of its own weight now seems remote.

My unpacking is done, and it is now almost evening, but I am not ready to consign myself to the trailer's interior for the night, where the plant's roar seems to reverberate in the sheet-metal walls. I walk up to the north end of Burtco Court and out into the old Northern Pacific spoil piles. I walk through the golden late-summer grass on the edge of a field that a man whose family homesteaded here has recently sold to Western Energy. I dodge a hole where the earth, undermined by slippage of coal beneath into the old open pit, has sunk ten feet. I climb to the top of a miniature volcano of spoil. Today, newly returned from the world of occasional noise, I am much more sensitive to the thunder of the power plant, and I notice that it is no quieter here.

I look out over the pit and past it, where five rows of spoil march out, showing bright edges fringed with thin grass, to the low sun. Past them is a glitter of bottles and cans, two pickup trucks, and the movement of figures: the Colstrip dump. Past that a dragline slowly turns, like a slow-motion dancer, flinging its arm out across the land. Today I drove in past an area on which the dragline had been munching last spring; now it was a long, smooth hump, like an English down imported to lie among the ragged hills of eastern Montana. There were bulldozers on its gray shale surface applying topsoil, which looked brown and earthy; although I fought the idea mightily, it was irresistible: It looked like a crew of undertakers applying makeup to a corpse.

I'm home. Down below me there is a sudden new noise, a sound like a small-caliber machine gun. In the shadows along the edge of the pit I see movement, and a plume of dust curls out of a ditch and gleams in the sun. In a moment the motorcycle rider appears in front of it, as if hurled up by a catapult, his wheels leaving the ground for a second, then returning to spout dust again. His helmet is white with a black visor; it is impossible to see his face. He leaps another ridge; the machine flashes in the sun; the noise rattles across the spoil. For a moment, in his own little world of sound and motion, the rider is freed from the tyranny of the plant.

59

Duke McRae seems sheepish about his kindness, as if it is a weakness, like temper in another man. It crowds up out of him when he can't afford it, a flare of understanding illuminating the other man's view when a narrow, angry light is what is needed. In eastern Montana, with the power plant roaring on the horizon, demanding action, tolerance is a liability. But Duke McRae is saddled with it—it permeates his being and spreads outward from him like a stain.

In the middle of a summer day I drive down the long dirt road to the big house near the creek. Duke welcomes me and leads me into the living room. He sits in a yellow flowered chair and I on a couch; he gives me two corned beef sandwiches made for his youngest son, Bruce, who is late for lunch because he is over at Colstrip registering for his junior year at high school. Sunlight drizzles in through shade trees planted the year the house was built, 1912, and casts moving speckles on the floor.

McRae is fifty years old. His smile is mild and unrestrained, his remaining hair is pulled forward to his brow, his shirt is unsnapped at the wrists, and when I talk too quietly he occasionally lets his hand drift up to cup around his ear.

"You lose your way of life, of course," he says. "In a sense we've completely lost community, as far as we're concerned. But when you get into the aspects of Three and Four, you don't know but what you might be losing *much* more than you can lay your finger on." Water, pollution's impact on trees and grass. "I think the unknowns are the things that probably . . . you know."

Unlike his cousin Wally, Duke McRae's eyes express no passion. Where Wally crackles with anger, real or manufactured for effect, lashing his rhetoric out like a whip to curl around an issue and sting,

Duke is like a powerful, slightly aging dog who knows he is playing with children and is more hurt than angry when the children turn cruel.

"[Recently] I realized that some of the coal under the land that I own was purchased as long ago as 1968 by Montana Power Company," he says. "Then they'd turned around and sold it to these other companies, like to Pacific Power and Light. So I went up to the Bureau of Land Management office in Billings and checked it out. And it had been sold. They'd come in and core-drilled it without any permission from me as long ago as '68, and sold it. Deeded land."

McRae's voice is patient, as if he was observing injustice from afar, not receiving it, and his sentences tend to trail off into unspoken understanding. He is so quiet that at first I think he is hardly interested, particularly in contrast to his wife.

In the large kitchen next to the room in which we sit, Pat McRae is working on some preparation for her fall semester with military efficiency, but she cannot help darting in to talk when McRae answers a question too vaguely for her taste. She is a perfect contrast to her husband: Slender, with a delicate but sharp-featured profile, she is quick, and seems wrapped in coils of energy. His shirt is loose at the cuffs, a comfortable plaid; she wears an immaculate pantsuit, gold earrings, a precision application of mascara, and a light scarf; her hair is tightly curled in a pale halo, and when she smiles it is sardonic, tainted with bitterness.

When I ask Duke what is the best example of what he has lost in terms of the community that he mentioned, he hesitates. "Well, I don't know," he says. "I think probably. . . ."

But Pat breaks in, suddenly appearing in the doorway.

"I could tell you!" she says, her voice sharp. "It used to be when you'd go to a ball game or to church or any community function, everybody knew everybody and it was all the old gang, and when Colstrip started developing there were all these strange people, strangers. And it wasn't bad until the strangers outnumbered the old-timers. You didn't feel too ill at ease as long as the people you knew were still in the majority. But when it got to the point where you went to a ball game and the old crowd was all scattered, then it was awkward. And over the last few years, the people who used to go to the ball games, whether they had kids in school or not, don't go anymore. Church is the same."

"I don't think it's so much the majority," Duke says slowly, "as . . .

as you lived in a community where there was never any animosity at all between the mining people that lived over there and the ranchers. . . . Never was for years and years. We all got along. But all this has completely gone. We still go to the ball games because we got two boys playing basketball, but. . . ." He searches his mind for the source of his irritation, which appears in Pat as apparent bitterness but in him as sadness. "They have a . . . what they call a 'Booster's Club' over there, and, you know, the boys play basketball. . . . But now, that Booster's Club's been going on I think three years but we've never been asked to join. And the reason they aren't asking me is because they don't feel you're a part of it."

His brow is furrowed as if, though it is a strain to articulate these things, he knows it is important enough to make the effort.

"And that's just the way I feel," he says. "I just don't think. . . . Possibly, maybe as years go on if they keep the same work force over there that maybe it'll be a community again, but it'll be several years. It'll be after . . . I say it'll be another generation."

Sometimes it seems that McRae almost worries for Montana Power Company, that its image should be so unnecessarily tarnished. There is none of the glee that Wally McRae sometimes takes in seeing the company step in its own public relations manure.

"I'm really appalled at the way a big company like Montana Power went about some of these things," he says. "I mean, their credibility couldn't have been worse." He leans back in the chair, shaking his head in gentle incredulity. "Of course, they'd had their way so long, they just felt there was no way they were . . . you know."

We talk about history—his family used to mine coal up where Peabody Coal Company is mining now—and we talk about Wally. I ask Duke if he is ever bothered by the controversy now attached to the name McRae largely because of his cousin's activism. He laughs, gently. There is no puncturing of the mildness.

"It doesn't bother me at all," he says. "We think enough alike that if they want to blame me for some of it, that's all right with me. But I'm not the spokesman that Wally is. I couldn't do it if I even . . . you know."

And we talk about the Northern Plains Resource Council's success in helping pass mined-land reclamation laws. "Why, when we first started, reclamation was just an unknown thing; when we started talking about topsoil that was the strangest thing that came down the pike." Now companies operating in Montana are required to stockpile the topsoil they remove from the overburden before stripping the land so they can replace it when the mining's finished.

"I think it's a shame to have to be forced to do all that stuff," he says. "It looks to me that those companies should have had the sense of responsibility on their own instead of . . . you know."

The irony, I think, watching McRae's face, is that today, when you tour any strip mine and reclamation site in Montana the stockpiling of topsoil is pointed to with pride as another example of the company's dedication to the philosophy of reclamation.

Finally we talk about Colstrip, and at last I think I am beginning to see the reality of his feeling. He is showing some irritation now, some anger in his voice and on his face. He sits up in the chair, elbows on his knees. Perhaps he trusts me enough now to let me see the frustration trapped in his life between layers of kindness.

We get into it backward, through the Forsyth meeting. Perhaps that is where the bitterness began, with Mike Hayworth assaulting him through the troubles of the past. "Of course I sometimes see Mike at a basketball game off at a distance," he says. "But I wouldn't talk to him under any circumstances. I mean, I don't have anything to say to him and I don't suppose he has anything to say to me so. . . ." Are there still hard feelings there? I ask. But he isn't willing to go to that extreme.

"Oh, I don't know," he says. "Not necessarily, not *any* feelings there. But I haven't anything in common with him, I don't think.

"But I don't go over to Colstrip at all now," he says, his sentences suddenly short, his voice crisp. "I don't feel I'm welcome, so I don't go. You don't have any input. You don't feel that it's your community."

And now he is fully angered. He leans forward, voice as hard as the scoria rock.

"Bruce is the only one over there at school now. He's got two years left. I hate it! I just hate every goddamn minute of it."

Less than a month later Bruce is dead.

60

The federal Environmental Protection Agency this week granted the request of the Northern Cheyenne Tribe to guarantee the cleanest possible air on the tribe's reservation. . . . The EPA decision to allow Class I air quality standards on the reservation may halt construction of Colstrip Units Three and Four. . . .

Allen Rowland, chairman of the . . . Tribal Council hailed the EPA decision saying, "There is no question in my mind that this spells bad news for the power companies trying to build those plants. . . ."

—Forsyth Independent
August 11, 1977

"I don't really think of my life as certain points," Marie Sanchez says, sitting behind her judge's desk in the tribal court in Lame Deer. Her speech is full of pauses, a little dreamy, like the flow of a big river that waits in eddies at bays. "I mean . . . I know your life . . . or the white people's life . . . everything is linear. . . . Here's where I graduated in 1957, here's where I did this. I suppose if I sat down and . . . wanted to have a linear account of my life like that . . . I suppose I could. . . . But I just kind of like . . . as each day comes, you know, . . . I just wake up. . . ."

Marie Sanchez is a tribal judge. Her office in Lame Deer is small. Just room for one desk and an extra chair. There are two things on the wall: a bright abstract of a bird in reds and yellows, and an American flag thumbtacked up behind her. She has a long face, long straight hair, and long slender hands. She is thirty-eight years old. She seems almost listless as she answers my questions, but whether this is a sign of genuine boredom or a smoke screen sent up by caution, I don't know.

When, I ask, did you decide to take a position against the power plant?

263

"I don't think it was really my decision," she says. "I guess maybe it was people like you. And now I really mean that. Even now I'm almost . . . I have to be careful of whatever statements I make. . . . Even my action . . . because anymore now they sensationalize everything I do or say. I don't know which reporter or writer labeled me militant. I tell you I don't have a militant bone in my body."

Don't you still belong to the American Indian Movement? I ask. "Yes."

Isn't it considered something of a militant organization?

"Well, it depends on who its opponents are."

Sanchez speaks softly, almost simply, but her eyes are wary under slightly heavy lids. She was never a leader in the fight for clean air; she testified at one or two hearings, but she was in the background. Like the Northern Cheyennes themselves in this conflict between miners and ranchers, she seems to stand a little apart; I'll never really know how important her role has been, if it is important at all. She is watching me take notes. She laughs gently.

"There was one article," she says, "where they had me standing out on the porch with—what was it?—a 30–30 rifle, saying 'There will be no strip mining today!' " She laughs again. "I *do* have a little knowledge about firearms. I'm a marksman . . . but I don't terrify coal companies . . . but I would if need be."

Back and forth. Activist, terrorist, teacher, judge. I cannot pin Marie Sanchez down. I ask her: You don't then eliminate violence as a means?

"No," she says. "I'm the great-great-granddaughter of Chief Little Wolf and so I think his spirit is still very much alive."

It is September in Lame Deer. The roads are dusty. Over by the tribal center in morning and evening when the cars and buses go to and from work and school the dust from the road rises to blanket half the town in a gritty mist. The town is mostly pastel-colored, federally funded houses, built in sections called Crazy Head Heights or Two Moons Subdivision, on roads named Little Wolf Drive, War Bonnet Drive, Antelope Street, and Little Eagle Road. Here and there small log shacks stand incongruously among the blue, pink, yellow, and green frame houses, and their very isolation makes them seem redolent with history. Sanchez's office is on the north edge of town on the road to Colstrip, not far from the large brick edifice that houses the Lame Deer office of the Bureau of Indian Affairs.

The Northern Cheyenne Reservation has been an official sanctuary of pure air for a month. It was officially given the Environmental Protection Agency designation as Class I on August 5, making it the first such place in the nation. On August 7 Congress passed amendments to the Clean Air Act placing the same classification on almost all national parks and wilderness areas. When the Northern Cheyennes learned of the designation, there was a general, if unorganized, celebration.

"We all took deep gulps of breath and we got hyperventilated," Marie Sanchez says. "And that was the biggest high we ever got."

But the Cheyenne desire for official clean air sometimes seems absurd. The Clean Air Act exempts minor forms of pollution from controls, and the Northern Cheyenne Reservation smolders with stove fires and trash fires, and the dust is so bad it contributes to respiratory disease. This apparent hypocrisy was one of the reasons Brother Ted Cramer felt the ranchers had more integrity in their cause than the Indians. "I think they should start picking up their own pop cans or beer cans on the road and cutting down on their trash burning in the dumps," he said once. And I remember now, while I listen to Sanchez, once standing in Lame Deer listening to a bright-eyed Cheyenne elder argue for clean air while the burning leaves in his yard contributed to a thin gray pall of dust and smoke that hung over the town. The thought is compelling that the clean air request was simply a political exercise, a reach for power, rather than any real yearning for environmental quality.

"The most important issue that I consider of great concern is survival of the Cheyenne as a nation," Sanchez says. Perhaps it's the political survival that matters, not the air.

And yet the Northern Cheyennes call the white man Spider. From a tribe to whom the most valued human qualities are wisdom, kindness, generosity, courage, and even temper, it is not necessarily a compliment. And the threat that the power plant brings is of the spider kind, an invisible web of tainted air. In the Cheyenne language the word for air is the same as the word for breath: *omotome*. And the air that sweeps down from Colstrip carries a strange breath of sulfur dioxide and very fine ash. The old Cheyenne I talked to near the fire of leaves applied the old Sweet Medicine prophecy to the plant, perhaps twisting it to suit: " 'He will put something up in the air that will throw out bad air,' he said. 'These things shall be far away but they will soon reach you.' " The old man laughed. "By God," he said, "that was a good one." So it may be that indeed the Northern Cheyennes do care with intensity about the air itself, it could be presumptuous of the white man to assume that the Indians must also weave lies and intrigue, though it would be naive not to expect cunning. But it is not the color of the air that they fear, not the stain on the face of the sky that blows away just like their own smoke and dust. It is the unknown infiltration of particles and gas from that incomprehensible machine; it is the slow float into their breath of an indetectable poison, like the silent tick of fallout; it is the inhuman dimension of that thin cloud that is so frightening.

My interview with Sanchez is ending. Her husband has arrived, a tall, very dark man with a small moustache. He sits in the corner, silent, wearing a very slight smile. In the end I ask her the question: What is so important to the Cheyennes about clean air? She, too, smiles softly, so I feel I am in a room filled with a delicate joke that only I cannot hear.

"Well," she says, "don't you like to breathe also? . . . I think it should be important to everybody. . . . Not just to the Northern Cheyenne and maybe not just to the state of Montana alone. . . . You shouldn't even ask us why we want clean air. . . . *You* know. . . . Other people through this whole world should want that."

As I leave her office, I notice that she has not just been doodling on her legal pad. She has written one word there. I don't know how it expresses the interview, or the conflict it was about. The word is "Bliss."

Her sleepy eyes watch me go.

But there is a spider in the works. "We will try to meet the Class I standards," a Montana Power Company spokesman says a few days later. "But if we don't meet them, we'll take the position we don't have to meet them, being grandfathered out of the legislation."

61

The shrieking has been going on for some time. It comes from a short vertical pipe in the middle of the power plant, which blasts out both an insignificant puff of steam and a tremendous howling noise. It's a pressure-relief valve, and the impression you get when it goes off is that it is set to open just seconds before the entire plant would otherwise explode and cover the landscape with salmon and green shards and bits of the human beings it keeps inside of it. As it is, the noise makes an effort to shatter eardrums. But it's like everything else around here: You get used to it.

The shrieking continues as I walk up past the plant toward the Colstrip Community Center, a new building that looks like a pair of sandstone monoliths lying tilted in the earth. The community center is, in some ways, a monument to Martin White's determination to make Colstrip a pleasant place to live, and to the death of the old town that was once here. In a corner of its spacious, sterile basement, where visiting tours are shown slides of mined-land reclamation, and high school students sometimes come to dance, is a shabby four-piece sectional couch, donated in the old western spirit of community by Wally McRae. It looks hopelessly out of place.

I go inside and the shrieking dies away in the general muffled roar. I have come down here today because this is the place Sheila McRae is employed.

It has been almost six months since I interviewed her in Billings and she vowed never to return to Colstrip. But now she's back. She's working as a receptionist here, arbitrating between the sweat of the basketball court and the chlorine of the pool. Her life has been a turmoil; she has been reconciled with Mike, who has been divorced and is living here in his parent's trailer; she has bought a twelve- by sixty-foot trailer of her own, and her brother, Bruce, has been killed in a single car accident on the way home from a party. Her smile, when she sees me approaching with notebook and tape recorder, is ironic and a bit challenging. Wouldn't you know you'd show up again, it says; you won't find me any different.

But she has changed. She is less formal now, no longer answering questions as if they were posed in front of a class. She now has a curious combination of languor and crispness in her manner; her words are bitten off short, as if to indicate determination, but she just glides down the slope of her ideas.

What brought you back? I ask, and she is as pleasantly vague as ever.

"Oh, I don't know. I don't know where else I'd rather live. Wherever there's work, and I have a job here now."

We don't talk about Mike. Instead we discuss a little incident in which she played a starring role.

Sheila was hired for the recreation center job not long before this interview. But at the same time, she says, she had a chance at a job as one of Martin White's secretaries in the Western Energy Colstrip office. At the time White warned her that the job would entail confidences and that she would come under fire from members of her family. Nevertheless, she said she would like the job, and she saw the recreation job as being temporary. Then one day one of the Western Energy employees came and told her that White had decided that the conflict of interest would put too much pressure on Sheila and that the job had been given to another.

"I had already decided that I didn't want it," Sheila says. "But I don't think he had any right to say that; that's being very discriminatory. I think they were more worried about my selling their secrets than they were about me.

"Just because my name is McRae!" she says, gnawing on a kernel of bitterness. "They're just backbiters over there. Everybody's so scared somebody's going to say something bad about them. They're just too worried about who they have working for them and who they're talking to and what they're saying."

But Sheila's words are shielded by a layer of soft amusement, a tone of irony and humor that muffles their reality. In this she is aided by another woman of about her age, fair-haired and flirtatious, who arrives in the middle of our talk and who seems amazed that Sheila condescends to talk to me at all. This woman's face can giggle slyly without making a sound, and she employs it fruitfully to try to push our conversation away from meaning. Together we banter along like a trio of bored sophisticates at a cocktail party nudging at communication with poles of blunt wit until, quite unexpectedly, we uncover the subject of Mike. We had been talking lightly of security, which Sheila defined as "having no bills, having a savings account, and being in a position to do things for other people." And I ask her, in the same light tone, if she would speak again at a hearing like the one at Forsyth.

"No," she says abruptly. "Definitely not."

Because you don't feel strongly enough?

The words come out suddenly shorn of their buoyancy.

"Because I feel strongly both ways."

There is a pause. The friend's face giggles blankly in the background.

"Because," she continues slowly, "I can see why my parents feel like they do and I can see why Wally feels like he does and I can see why Mike feels like he does, and why Mike's parents feel like they do. I can see everybody's point. I'm better off just to . . ."

She trails down into silence and I am reminded of Duke, sad and afraid of the pressures on his family's life. When Bruce died, and they played the bagpipes for him up on the big open field of the Lee Cemetery across the Rosebud road from his home, what did Duke hope for the future of the next generation, with Sheila and Pat beside him and the boy in the ground and the power plant grumbling just over the hill?

"Anyway," Sheila says, looking away between me and the friend, "I don't really think of Three and Four as one of the bigger parts of my life at all."

What are the bigger parts? I ask quietly, trying to keep it serious for just a moment more.

"My parents, my family. Mike."

Aren't they all tied in with Three and Four?

"I don't think of them that way." She is aloof again, but the defense of humor has not returned. "They all matter to me equally." There is another pause. Then the words come sharply, carved with precision. "If they didn't, I wouldn't be having any problem at all because I would be able to choose one over the other. And I'm not going to ever do that."

But to Sheila the rest of her future is impossible to see, and we drop again into farce. The friend is relieved. She starts passing around cream-filled, sugar-powdered doughnuts. But I suddenly think of Sheila as a boat, moored by training in what was once a calm sea. And then the power plant came, and either because she was loosely anchored in tides that had always been kind, or because it was a huge storm, it cast her adrift. And as we continue in a conversation now almost desperately leavened, I see her spinning in the waves. The future? Move on, move on. Her friend munches and mentions that nobody likes Colstrip, anyway. Sheila nods vaguely, glancing out the window as if saying that it's time for this interview to end. Maybe the next place she goes will be better, I think, or so she hopes.

"I won't be here in two years," she says without preamble. "When I get rich I'm going to leave."

But you won't get rich working for the Colstrip recreation department, I say.

"Then maybe I'll marry somebody rich and he'll take me away."

But then, I say, if you marry somebody who has become rich working at Colstrip you'll just go from Colstrip to Colstrip, all the places he works, all the rest of your life.

She looks out the window, at the green park where at night kids in four-wheel-drive pickups tear up the grass, at the dragline swinging patiently back and forth on the ridge. Her profile is haughty, her eyebrows are slightly raised in pale brown slashes of defiance. Her friend, sardonic as a cat, licks her fingers of sugar and waits for the answer, amused.

"A lot of people do it," Sheila says.

62

More than one-third of a billion dollars will be on the line Tuesday afternoon at 2 P.M. as Rosebud County commissioners hold a public hearing before deciding to go ahead with the issuance of industrial revenue bonds to help finance pollution control equipment on Colstrip units Three and Four.

—*Forsyth Independent*
September 12, 1977

On our way to the Wimer ranch—Wally McRae and I in his cattle truck, driving up the long dirt road on which Charlie Wallace was killed—we come to a ridge, and here, silhouetted against the blue distance of the hills on the other side of the Rosebud are two people on horseback. It is a picture out of the Old West, out of the Charles Russell museum in Helena: the early sun, the riding shapes, the cows moving before them in the bright-lit prairie grass. McRae suddenly repeats, from memory, the speech he often uses at the beginning of a rodeo as the contestants ride together into the arena:

"Before you, you see cowboys and cowgirls by the score, suntanned sons and daughters of the Golden West. They, and others like them from the shores of the Pacific to the farms of the Nation's heartland, from the prairie and mountain provinces, deep into Old Mexico. These hardy individuals with the clear eye, and the steady nerves. . . ."

His voice trails off. Go on, I say. He shakes his head. He seems to have stumbled over something—maybe his cynicism; maybe his thought of the strange arena of controversy in which he now performs—next week he will be in it again—or maybe a memory of Charlie Wallace, the rodeo bull rider, who died here just a year ago when his pickup went off the road.

"Ah," McRae says finally. "It says everything and it says noth-ing." The truck rattles up over the ridge, and I think: It is the glamour and the glory of being a cowboy that he is so cynical about acknowl-edging, but it is the glamour and the glory that he loves, too, the truth in those trite words, that cowboys still do ride, chapped and hatted, against the early-morning sun on the bright grass hills of the prairie, and that theirs is still a special world.

The truck spins dust up behind it. We meet the two riders at a corral in a cup of hills near the Wimers' old ranch house. They turn out to be Tom Wimer and his daughter, Charlie's widow, Barbara.

From several places on the Wimer ranch you can see the plume of steam from the power plant, visible in the cool morning air; preoccu-pied as I am with the struggle over Three and Four, I find it strange that here it is totally ignored. The life of the ranch continues without mention of the plant. But they don't mention Charlie, either.

The day is dry. By noon it will be hot. All summer the weather has been like this. The grass has parched under the sun's unbroken stare. The cattle have mowed it down to nothing too quickly, too soon. Thunderheads have formed in afternoon skies, then withered, perhaps speckling the roads with little splashes that cake in the dust and are not washed away. The fickle rain has been gone too long and now the ranchers are beginning to pay.

Tom Wimer is small and stocky, and today he looks worried. Bar-bara is quietly cheerful, a serene woman, in spite of what the day will bring. Like McRae, and like a couple of other men who have brought cattle trucks to help today, they wear the old wide-brimmed hat of the West. It is a special day, and the hat is the dress uniform: Today the Wimers are selling their season's calves. Wimer doesn't think he has enough grass to keep the calves another month, so he's selling them early. By the end of the day he will know just how much this caution has cost.

The Wimer family, which is made up of Barbara Wallace; the son she bore Charlie just after his death, who bears his name; Tom and Donna Wimer; and another daughter and son-in-law, has, like most families here, long roots in Rosebud County. Wimer's maternal grand-parents homesteaded here in 1910; his father homesteaded here in 1916, and he was born here in 1919, just six years before Northern Pa-cific opened its mine and created the town of Colstrip. But though they can see the power plant's plume from the window of the new house they built on a ridge above Armell's Creek, they live on the edge of the controversy, reluctant to become a part. Barbara is the only one who seems to take more than an observer's interest: "I don't know whether I would send Charlie Bill to school in Colstrip or not. If the plants go through I don't think it would be a very good environment." But Wimer himself is cautious. "If the impact of the people doesn't get any worse than it did when they were building One and Two, then I would say we have no complaints," Tom Wimer told me once. "We've been lucky and the people have been good."

After we arrive at the ranch the pace accelerates. The air is laden with tension; McRae, Wimer, Barbara, and Jack, Wimer's son-in-law, work fast in the deep manure. There is a war on between the cattle buyer and Wimer here over the matter of "shrink." Wimer wants his calves to have a full stomach when they step on the community scale down at the stockyards on the Rosebud; the buyer wants them empty. The difference can be 5 percent. So there is a race with nature to get the animals sorted out from their mothers and up the chute into the trucks, and down to the stockyards before the calves eliminate some of their weight. It has been said that this is one of the few jobs in the world in which you get paid for shit.

The air is full of noise as the calves bawl and the cows, already getting fat with next year's crop, return the music. Wimer rides around in the corral on horseback, overseeing the operation, casually urging calves into the chute when he can, looking a bit bewildered in his worry that he is selling too early. Barbara Wallace stands at a spot where the calves must be turned and waves her arms up and down like a child making windmills in snow. Wally McRae flicks his short whip vigorously, again so charged with energy that he seems to work angry. Above it all, from the roof of a small stable, a goat watches placidly, occasionally scratching its neck on the electric power line that stretches out to a light in the corral.

No one mentions the name of Charlie Wallace. He is cleanly gone, the young man who so joyfully accepted both the romance and the reality of the cowboy life. Yet even in his death he has become a kind of symbol of the strange turmoil that has shaken this county. Wallace is buried in the Lee Cemetery across the Rosebud road from Duke McRae's home. Everyone knew that is where he would prefer to be; even his parents in Kansas agreed. Yet the acceptance he earned in four years is not offered to or requested by most of the people who now live in Colstrip. They don't want it.

"Probably the hardest thing I found at work," Barbara Wallace remembered once, "was that every day you'd hear people talking about 'I hate this place, I can't stand it, and I wish I wasn't here.' It was just about everybody. They were just here to make a buck and get out. Western Energy was doing a nice job on the streets and the parks, and trying to make a model city, but I think they have forgot that people make the community." There is no one like Charlie Wallace in Colstrip. And there is no cemetery in Colstrip, because no one there thinks of Colstrip as home.

McRae's truck is loaded and we leap in. He takes off down the road, the moving cattle in the back changing the balance of the truck so that it sways along, as if in a wind. We say nothing as the truck roars past Colstrip like a spy strolling through the enemy camp. I think about Barbara Wallace coming down here to work in the Bechtel office every day after making that speech in Forsyth. "Some people were sure I was going to lose my job," she told me. "I wasn't particularly worried about it; I didn't figure they'd have the guts to do that. But I suppose for the next month, every time someone came by my desk they'd have to stop and tell me about Three and Four."

I remember them talking up in their new house on the ridge, Tom and Barbara. "I kind of wished I had about eighty feet of coal over the whole outfit," Wimer had said, "and I could just let them have it and go someplace else, that's what. If the country's gonna change so much you don't want to live here, you just as well have something so they can pay you for moving. But as far as we know we don't."

And Barbara had replied:

"The thing I say is, okay, you feel threatened here, you move, okay, so in a few years you're going to feel threatened somewhere else. How long can you keep moving? Maybe we better just fight for saving this right now."

The rancher's year's end plays out under high mackerel clouds at the Rosebud stockyards. The calves are unloaded and sorted into pens. The buyer strides around, eyeing them carefully, grinning, his vast belly billowing out a pink shirt. The calves are moved in small groups to the silver-painted scale, and they stand on the jiggling platform all spread-legged and wide-eyed, like children on a train. When the weighing is done the buyer gathers the papers together, rests his belly on the hood of his big green Pontiac, and concludes the sale. Around him huddle the cheerful veterinarian, the lean and bony brand inspector, Wimer's other daughter, Carol, and Wimer himself, trying not to look tragic. All sign. The year's work—the haying, the feeding, the calving, the branding, is signed away on the hood of the Pontiac. The rest of us stand around eating Donna Wimer's cherry pie and cookies.

"I stopped baking two years ago. This is my first since then," Donna says.

"Wouldn't know it," says Wally McRae.

Behind us, the last part of the little drama ends. The calves are herded into the cattle truck. The huge tractor-trailer rig stands waiting, a silver cage stained with manure. The driver wields a long electric prod. He wears greasy overalls, a two-day beard, and a baseball cap; he is not of the fraternity on this special day. "Heyyy—Ah!" he shouts. "Whoooop! Hey, hey, hey, hey." The calves clang on up the ramp, scuffle in the dark interior. Their shapes cover the spaces in the steel latticework; the calves are like prisoners behind a board fence—all you see is shadowy movement and an occasional eye.

Wimer walks away from the Pontiac. The little group disperses. Wimer's shirt is stained with sweat and coffee. The middle finger of his right hand bleeds around the nail. There is manure on his trouser legs.

"They weighed awful light," he says, to no one. "We had grass in the winter pasture. Should'a used it. If the rain'd come. . . . I bet I lost five, seven thousand dollars."

"Hush, Tom," Donna says.

The small group collectively sighs. The race with shrink is over. The urine and manure the calves now spread on the metal floors of the big trailer has been sold. The stock trucks pull' away one by one and clatter back up onto the paved roads, their slatted racks rattling, empty. The little building that houses the scale is locked until next time. The buyer gets back in his Pontiac, ducks his head low to avoid hitting the crown of his big hat, grins, and drives away. And for a few hours the power plant has been invisible. The old routine of the cattleman's life has expanded to fill all space: the roundup, the riding, the shouting, the weighing, the anxiety of decision. But the boy to whom Duke McRae handed on that oral tradition, like a baton, is dead, and just up over the hill the power plant roars like a steam engine headed west. Tomorrow will be different.

And indeed it is. It rains.

63

On September 19, 1977, two years and ten months after the people of Rosebud County came to Forsyth in a snowstorm to do battle over Colstrip Units Three and Four, another public meeting is held in the county courthouse, for the same purpose. Very little has changed, except the specific function of the meeting, which doesn't really matter.

This matter was called by the Rosebud County Commissioners to discuss the five companies' request for the issuance of about $350 million of industrial revenue bonds—those same bonds the Rosebud Protective Association had discussed opposing eight months before. The bonds would save the consortium about $80 million.

I ride down to the hearing with Wally McRae, who uses the trip to Forsyth to deliver a load of calves to the city stockyards there. As he drives into town in the big green stock truck, towering over the other traffic, a hurrying group of about six men in suits and ties blunders out into an intersection in front of him. They are the Montana Power Company contingent, also on the way to the hearing, a phalanx of power on the dusty road, clad in blue and gray, carrying briefcases and leaning forward in unified haste. Coattails flap in the breeze. With no change of expression, McRae lets the truck roll right up to the edge of the hustling group, then slams the brakes so it shakes and rattles to a stop.

The little group scatters, its leaders dashing for the far sidewalk, the men behind leaping back, uncertain of the truck's course. The phalanx breaks into disarray. For a moment its members stand still, struck apart. Then McRae grins and sticks his head out the window so they can see who it is. The men smile back and wave, a little uncomfortably.

"I'm missing a golden opportunity," McRae shouts to them, and lets them all pass and rebuild their formation before he drives on. But it would not be like McRae to ignore the symbolism of the event.

Inside the courtroom the hearing is full of echoes: The room is crowded, and though there is no chanting, there is restlessness in the back and people are talking in the hall outside. The commissioners sit in front, three men, among them Duke's brother Mac McRae, who sucks on a cold pipe. They watch a familiar parade—same old majorettes, same old band.

There are speeches by Wally McRae and Don Bailey arguing that the companies are capable of removing 90 percent of the sulfur from the emissions instead of the 40 percent the companies promise, and there are speeches in return saying such equipment is impossible to buy. A Montana Power Company official commits the error of admitting that the plants might die if the bonds aren't approved, then gets up and withdraws that statement after consultation with his superiors. And Vic Jungers is there, jutting his angular face out at the crowd and advising the commissioners like a consultant to take a long look at the overall future of the county before making their decision. But the rest of the people who had spoken at Forsyth in 1974 are either not here or are hidden in the crowd. I spot Duke McRae sitting in a row of ranchers halfway back, but I see none of the others. The years have brought attrition to the ranks of both sides, and new faces fill the gaps. There is a new stridency, though, in the replacements, even less of a sense of taste, propriety, that Wally McRae seems to cherish.

Ted Cramer's place as the representative of the Lord, for instance, is taken by Andrew Goodwin, pastor of the Colstrip First Baptist Church, who doesn't just suggest by his costume a holy presence.

"The more people who move to Colstrip," he says, "the more people we can reach for Christ. This will help the growth of churches in Colstrip. A lot of prayer has gone up over this and I believe it's in the will of God. And it's no use fighting against something that's in the will of God."

And the role of demagogue, which had not even been taken at Forsyth in 1974, is ably filled by a Colstrip milk dealer named Ray Loveridge, who, having everything to gain by the construction of Three and Four, demands sacrifice.

"We have out in front of our courthouse a monument dedicated to the men who have served in the armed forces of this country," he booms. "And the bottom line says sacrifice. I believe, gentlemen, we'd better compromise and sacrifice."

McRae leaves the hearing late in the afternoon. The resonance between the two meetings rings in my head and there is a sense of timelessness whirling like leaves in the early fall air. And I think about something Kit Muller, the NPRC staffer, had said a year before. In a wry moment he had proposed a bumper-sticker slogan that could be used with accuracy and vigorous partisanship by all sides. The sticker would say simply:

COLSTRIP IS FOREVER!

On the way back to Colstrip McRae is silent for a long time. Again I am aware of poised events; this has become commonplace in the years I have spent with Colstrip—you are always wondering if this decision will be the last. And I know that McRae is depressed about the meeting; about the knowledge that Ray Loveridge, a man whom McRae could not respect, was being groomed by Martin White as a Colstrip leader, possibly even the town's first mayor; about the new distance that has grown between him and White. "I'm going to quit talking to Wally about this stuff," White told me a few days before. "I don't want to get into arguments with Wally. I'm going to quit that for a while." Earlier, McRae had said: "It is significant that Martin hasn't called me lately. I had thought there was such a thing as having different points of view and having community compatibility—and apparently I've been manipulated, too."

Now we talk about what will happen if Three and Four die on the spear of unissued revenue bonds, which neither of us expects. Again McRae does not want to talk about winning, but this time his mood is sad, not conciliatory.

"If Three and Four are not built," he says, "it's not a victory. There shouldn't be a celebration. It's been too bitter. It's been too rough. The community has been destroyed."

The aftermath of this hearing is grotesque. The commissioners vote to approve the bonds under the condition that the companies install pollution control equipment that will be guaranteed to remove 85 percent of the sulfur from the power plant's emissions. The companies turn to the Forsyth City Council, a group which has no jurisdiction whatsoever over Colstrip. "I am appearing here with my hat in my hand," says George O'Connor, former president of Montana Power Company. "I haven't got any other place to go." The city council agrees to issue the bonds, for a price. The companies will pay the city about $735,000, and pay the Forsyth city attorney about $35,000. "There's always a little mystery about a prostitute," Mac McRae tells one of the council members. "But that's been cleared up now; we know what your price is."

64

Caught between the U.S. Environmental Protection Agency and the Montana Power Company, more than a hundred angry union members protested the work stoppage at Colstrip Three and Four Tuesday by picketing outside the Federal Building in Billings.

[They] carried signs that read, "Starvation Kills Faster Than Bad Air," and "Out of Work and Hungry? Eat an Environmentalist!"

On Monday, about 100 skilled workers were idled when Montana Power complied with an EPA order and halted work on the two new generating plants in Rosebud County. . . .

Ranchers on the Northern Plains Resource Council . . . say that Montana Power is using unfair tactics by pitting the workers against ranchers, farmers, and environmentalists.

"They're pawns in a bluffing game that the power companies are playing," said Don Bailey. . . .

He said Montana Power should not have begun hiring workers for a plant they never had the right to build.

—*Billings Gazette*
October 13, 1977

In the cool, damp aftermath of a storm, I walk out on Wally McRae's land. I ask permission at the ranch house and then walk straight out, past the row of junipers planted against the wind, and down across a draw, where there is a cluster of cottonwoods and a trickle of sand. And I slowly climb the grass- and sagebrush-covered hills. The land is gnarled; here there is a low cone of white and crusted mineral like salt, an alkali pimple on the land; here there is a knob of scoria; here there is a short sweep of grassland that looks smooth but as I walk across it

280

suddenly reveals a gash made by spring meltwater and summer rains. There is white rime in the gash, too. I cross a fence at a corner, holding the wire down and swinging over well clear of the barbs; here there is one of those many sandstone rocks you find out on the big ranches, carved in a strange design. McRae tells a story of a city woman who made a hobby of collecting these stones, believing them to be Indian artifacts, until one day a rancher told her their real purpose: to mark the surveyed corners of the open land.

I walk across a flat ridge and down a little slope of brown grass; at the bottom, the growth all blasted away from it as if it were a flower of death, is a salt block, carved into sensuous curves by the caresses of a hundred tongues. The cows are well spread today; they are mostly far from me in little blobs of black and red and white against the brown; when I approach one she snorts and canters away with a combination of weight and bounce, like a bloated deer, until at a more convenient range she can look back over her shoulder with a glance that is hardly coy.

I walk and walk. The hills climb slowly, reaching toward a long bluff ridged by pines; from this open country, where I stand higher than everything but the hills themselves, the trees seem somber and dense in the distance; they do not invite me into their shadow in this waning day. Today I prefer the naked country, where, every once in a while, when you do not force contemplation, the spaciousness springs you from the trap of thought with a burst of intensity and your eyes and mind flash into distance and away like sudden birds. Your eyes peck at the light, the prickles of buffalo grass nip your fingers, there are wild burrs of laughter in your hair. The land crowds you, it leaps away: The red ridges, the black, bare trees in the draws, the water like bitter syrup in the roots, the dry, stinging scent of the brush, the ranks of cloud, so still, so swiftly moving.

I climb a little hill, thirsting for that release that I have found here before, but when I get to the top, puffing, the distance is just distance, faintly hazed. So I play with the bits of scoria on the top, throwing handfuls down the slope and listening to the tinkle they make. A meadowlark sings, invisibly, nearby, and I am reminded of the testimony of Patricia McRae.

The earth at the top of my knoll is stony but soft; I break off stems of dry grass and jab them into it like needles into skin. The cattle honk down in the little valleys. It is still, and across the creek beyond the ridges I can see a pillar of steam standing upright in the windless sky. Pillar of smoke, pillar of fire. Out here, where nothing is being consumed but the grass and time, I am suddenly oppressed by the notion that in that white column is packed all our roaring society of machines and light to which we are so willingly enslaved: sheets of glowing steel; the clang and glitter of Las Vegas; motel lights strung along freeways; telephones ringing, clicking, buzzing with friendship; printing presses rolling out slabs of news; furnaces breathing heavily in empty buildings; high glass and air conditioners; motorboats; jets slung like flakes of phosphorus across the sky; freeways of cars pouring in belts of steel under heavy skies, rolling contained together all still to one another as on the day they were born; microwave ovens; televisions like blue-green tanks of light swimming in darkened rooms beyond conversation; automatic staplers; aluminum plants making aircraft out of vats of metal soft as margarine; coffeepots, copiers; cold rooms where meat is hung—

I leave the little hill on McRae's land and walk down the steep scoria slope into the valley. Chips of brittle rock tinkle down the hill ahead of me in little cascades. At the bottom cows are calling to wandering calves across the dry grass. As I descend the plant diminishes behind the hills and behind the pine trees on the ridges and finally disappears altogether.

65

It is a Sunday morning. I am standing in the little alcove in my trailer that serves as a makeshift office, talking to my wife at our home in Idaho, when a girl opens the trailer door and walks in. I hear the sound of the plant go suddenly loud. I look around. She is about five years old. She closes the door carefully behind her; the roar subsides to its normal background thunder. She looks at me across the room. She has big brown eyes, blond hair, and dirt around her mouth. She stands waiting expectantly, as if she has brought a message, or expects one. I've never seen her before. She stands on the warped green linoleum squares in the kitchen and watches me as I listen to letters that have accumulated at home. I say nothing. The girl says nothing. Her eyes are not amused, unhappy, eager, or curious. They are barely alive. The noise of the plant washes around the trailer like wind. My wife seems to be talking from another world, a strange place where there are trees, rivers, and silence to be had just for listening; a place in the past. I turn to my desk to write down a telephone number, and when I look up a moment later the girl is gone. I go to the door and look outside. Acres of trailers. No faces in the windows, no children on the stony common ground. She has vanished. I go out, lock the door, and go to church.

For all its wealth, the St. Labre Indian Mission always seems strangely abandoned. Sometimes, when you come upon it in surprise on the edge of the reservation town of Ashland, it looks like a deserted settlement in all its stone and stucco elegance, left in a rush to the shy natives by some technologically superior race that had no time to burn it in its haste to get away.

Once I stopped at the mission in early evening on my way to Colstrip, and I wandered through the grounds looking for a telephone. I

283

walked on dry lawns and concrete sidewalks between the acres of low buildings to the base of the stone tipi that stands like a monolith at the complex's center. No one else was there. There were no faces in the windows, no distant music, no drums, no choirs. It seemed to be a ghost camp; I half expected to find, when I probed the open doors, blackboards inscribed with last year's homework and tables set for breakfast with old food never eaten.

"You always wonder what happens there," Wally McRae told me one day. "You think it's something important, and you tell yourself that you're going to find out. But you never do, and you keep driving by and you get used to it."

Today, though, the huge stone tipi, which is the chapel, is over half full, although the worshipers are mostly white. The outside of the building is a monument to stone masonry; the interior seems weak. The power of the outer banks of stone is gone—the inner walls are white and blank; the chairs are soft and softly colored in red, green, beige, and orange; the floor is pebbled concrete. The three shafts of stained glass in the roof shed an ineffective light: It is bright enough, but the morning sun hurls its brilliance high on the wall and seems remote and unattainable to the mortals on the floor. The room is tall and spacious indeed, and you feel small in it, but you seem diminished not by the grandeur of God or nature, as you do beside a mountain, but by the skill of engineers. In that sense, the chapel is a little like the power plant.

There is a sign on an easel beside the raised circular altar which says, on green cardboard, ALL I ASK OF YOU IS FOREVER TO REMEMBER ME AS LOVING YOU. Near it three priests stand facing the audience, which is divided roughly into two-thirds white people and one-third Indians. The Indians sit together to the right of the altar. Near the three priests is a quintet—four guitars and a double bass. At the bass is Brother Ted Cramer.

Cramer is a lean and somber man. Today he wears the Capuchin hooded robe and sandals; Friday, I had found him in a turtleneck and slacks mixing sound in his den of electronics on the other side of the campus. There we had talked about his flying and the way he had been drawn into the controversy over the power plant, then he had invited me to sit in on a religion class for elementary school students. The subject of the day was respect; here too there had been signs on the walls: ALL I ASK OF YOU IS FOREVER TO REMEMBER ME AS LOVING YOU. But the class had been disrupted by an Indian girl who became upset at an obscure insult made by another child and whom Cramer had to take out of the room to counsel.

Afterward we returned to his sound room and there he had vented the frustration perhaps caused by his mission's reluctance to get fully involved in what he saw as a massive injustice.

"It seems to me the government should be supplying them [opponents of the power plant] with funds to fight the companies," he said, then, with no extra weight to his voice, added: "I think they should force that in any way possible—violence if need be. And I'm a little against violence, but I'm beginning to think there isn't much choice."

After the class, the words had seemed out of place. And Cramer himself seemed pale and small behind their thrust. What sort of violence? I asked.

"Strong. I would be willing to be hired by the power company and work myself up in there so if I—if I could get into one of those gears, I'd throw something in there. I would be willing to supply dynamite."

But today, Cramer no longer seems pale and powerless. It is as if the habit of the monk—the sweeping dark brown robe, the white rope belt, the hood hanging behind—has combined with the deep power of the instrument he plays to draw out a ferocity. Cramer's moustache is no longer just a foil for a weak chin. It is belligerence. I lean back in my comfortable chair. The quintet is singing a gentle song: "Lord, give me a heart of flesh, to hear your word, your love; I've had a heart of stone for so long." Three small Indian boys are fidgeting in the front row of the right section of chairs. I permit myself a fantasy, built on Cramer and the future and the tour I took yesterday into the interior of the power plant.

In the dark the power plant bellows. It stands out in the night like four ships lashed together in their own storm. Colstrip Units One, Two, Three, and Four have all been on line now for almost a year, spewing power out across the country. In the dark the lights around the plant throw their colors against the rising steam and the whole complex seems to be rising and curling in the green and amber turbulence. The chain-link fences that surround the plant glitter with moisture; high in the cloud the strobe lights flash, outlining the racing steam.

Inside the plant routine is as comfortably installed as it can be in a place so devoted, it seems, to the production of catastrophic noise. On the ground floor, near a door, the concrete floor is stained white; there is a smell of sulfur, and the ceiling is layer upon layer of pipes. Over the general roar there is a sporadic and local hissing, and vagrant clouds of steam rise like ghosts from vents in the floor. Trickles of water gleam in the tangled hallways.

It does not take many men to run the night shift. Hard-hatted creatures prowl the corridors on the upper stories' red-painted floors, maintaining the flow of power. They are indistinct in the tangle.

The coal's travel to the fire is invisible. In its powdered form it is blown, rather than carried, by fans the size of houses, in a curving trajectory that leads it to portholes in the boiler walls. The fans are responsible for much of the interior noise, as they pump the tons of black powder up from the silos.

In the control room three men examine curving panels arrayed in orderly patterns of switches. On the wall is one touch of humanity—if that's what it is. Someone has hung up a photograph of a powerful motorcycle flinging mud toward the camera. Nearby on the panels are displays of warning lights; as usual, a half dozen or more of these are red, but the men seem to ignore them. High on the panels are four closed-circuit television screens, which show a pale, formless picture of the fire.

In the four boilers the flame rages, as if the men in the control room had temporarily trapped it in a jail cell. The coal bursting in explodes before it settles, and the interior is a storm. The portholes in the inspection doors high on the boilers' walls gleam with red light, and visible through them are hanging banks of steel pipe. In the blast within, the pipes swing stiffly back and forth like frozen laundry in a wind. Back in about 1975—fifteen years ago—back when the relationships had not deteriorated among residents of Rosebud County, a Montana Power Company engineer had given the notorious Wally McRae a tour of the plant. When they got to the portholes and the engineer opened one to show McRae the blazing eye of the fire, a few hard sparks had flown out, like brimstone. The engineer had leaned up to McRae's ear and shouted: "That's where you're going!"

The story's not told much anymore.

In the vault of the generation rooms, the machines that are the heart of the plant spin their web of power. Clad in half shells of yellow-painted steel, the generators lie sunken in the floor like logs embedded in sand. Pipes bearing lubricating oil rise to feed them, and then drop away to remove the spent dregs. At one end the steam lines deliver their thrust, and at the other more pipes take steam drained of strength almost to the point of water, for a trip to the cooling banks and then return. It is as if the incoming pipes come offering gifts, which the generator devours and discards.

The power that leaves the generators is taken away beneath the floor, as if it was body waste. Under the rumbling concrete the conduit flashes the white liquid power out to the switching yards, where it is lifted into the wires and hurled west, to Billings, Spokane, or Seattle. In contrast to the plant, which shouts most vigorously of its creative travail, the power lines only buzz faintly, as if they are carrying a terrible secret in their swings across the land.

Inside the plant one man on the night shift has diverted from his routine of inspection. He is not a large man, and he shaved off his moustache years ago, when the second two plants were approved and the desperation of another construction season descended on Colstrip. In that grim spring, when Don Bailey sold his outfit and left for Alaska and the Rosebud Protective Association dissolved in defeat, this man had taken a job with Montana Power Company as a janitor and had slowly worked his way up to this position of responsibility, a nightly inspection of the interior of the plants. Now, with the pit for Units Five and Six yawning daily wider to the east, his time had come at last. Martial law had descended twice on Rosebud County in the past year when rumors of a Northern Cheyenne attack on the plant had been carried to the federal energy department; four Indians had been shot on the Northern Cheyenne Reservation during the National Guard enforcement of court orders giving eminent domain to the companies who wanted to mine the coal beneath it for their synthetic gasoline plants at Birney, Ashland, and Hardin under the so-called 1982 Energy Supremacy Act; and a rancher who had been speaking at the two-day-long hearings on the Colstrip coal-gasification plant was seriously injured by a runaway bulldozer during a tour of reclamation sites; but everyone had forgotten the former Capuchin monk.

And now there was serious talk of setting aside "Energy Sanctuaries" in which political freedoms in communities that served huge power plants would be slightly curtailed to prevent the kind of terrorism which had shut down two cities during the past year when five members of an obscure underground cult had damaged the central-station power plants on which the cities depended. The argument was that benevolent but strict military rule of a small number of people in a few remote parts of the country would help ensure freedom for the rest of the nation. Rosebud County was on the list. So the man who now sets the conclusion of his long plan into action knows he won't get another chance.

Inside the plant the noise of its operation is occasionally over-powered by the croak of the loudspeakers. The small man cannot help thinking as he has before that the sound is like the voice of a brain summoning muscles. The thought makes him bitter; it seems too true. He quickens his pace. Every night he makes this same tour, flashing the powerful beam of his light up into the shadowed places. But tonight he is carrying a toolbox around with him.

This is not unusual—the man has made himself handy around the plant—but tonight the box contains, instead of wrenches, a stack of small packages, each with a tiny antenna protruding from its body. The man knows exactly where to put them; he has spent his spare time during the past year in meticulous study of both the plant's points of weakness and of the use of explosives. Now he plants each package in a different location, clamping it securely on a pipe or tank, one of which reads, in stenciled letters designed for the tourists who pass through the place daily, 4,000 GALLONS LUBRICATING OIL. (Nearby, against a steel bulkhead, is a similar sign reading EFFLUENT PUMP. (Years ago some-one had painted, beneath *Effluent,* the word *Society.* Such levity in the plant is no longer tolerated.)

The small man's operation takes two hours. He passes through each vast building, and pays a visit to each turbine, each vat of oil, each line of cooling water. Finally he walks wearily, and a little reluc-tantly, to a room where he has a locker. In this locker he places the toolbox, now nearly empty, and he wires a larger antenna on its lid. He pauses. A flicker of sadness crosses his face then disappears. He turns a switch on the toolbox. It hums.

The man runs for a telephone nearby, pushes two buttons on its box, and lets his voice boom out into the plants.

"Emergency. Evacuate immediately. Emergency. Evacuate imme-diately. Evacuate immediately. Evacuate immediately. You have ninety seconds. Evacuate immediately. You have eighty seconds. . . ."

In a minute and a half the place is empty. Only the man at the telephone remains, still shouting into it, when the humming stops in the toolbox and the roar of the plants is punctured gently by a series of explosions that seem so mild they can hardly be heard outside. One of them silences the man at the telephone by blasting a gush of hot oil over him, and knocking him to the floor. The interior of the plant fills quickly with steam, and becomes too hot for life.

In the darkness the plant begins to scream. The explosions were smothered by the general noise, but this shriek, the last call of all pressure-relief valves in the structure, is transcendent. It wakes the town. Martin White leaps into his car and drives to the office, where he gets on the phone to Butte. They already know; lights are going out all over the Northwest. Lois Olmstead sits up in bed and prays. Vic Jungers quiets the dogs and stands at the doorway of his trailer, curious, half amused by the noise. Myron Brien curses and rolls over. And down on the Rosebud, Wally McRae hears it and gets up slowly, instinctively aware that the great beast is in pain. He stands at the door, chewing the soft end of a match.

The shrieking lasts until dawn, slowly diminishing as the pressure dies and the boilers cool, crackling and sighing. The generators lie frozen with the heat, their yellow fairings curled and blackened, the paint frayed. Inside they are welded steel; fans a part of housing, shaft forever still.

And as the sun rises on Colstrip, a crowd gathers outside the chain-link fence, looking up at the four stacks, the four structures, the soiled salmon and green and red and silver of Colstrip One, Two, Three, and Four. They have all come to see it, the heavy equipment operators, the engineers, the janitors, the teachers, the chemists, the maintenance men, the computer technicians, all have come to see the machine that they have depended on, that has ruled them, for years. No steam escapes from the steam plant. It is dead. And there is an odd thing in the air that some of the children of this town have never experienced at home until today: silence.

In the chapel of the St. Labre Indian Mission, the quintet is singing. The monk in the brown habit plucks the double bass. On the wall is a slogan: MAY THE GREAT SPIRIT MAKE SUNRISE IN YOUR HEART. The song is a kind one—a gentle Christ, a human spirit. Ted Cramer sings with all the rest: "All I ask of you is forever to remember me as loving you."

Part 4

66

The power plant has stopped! It is quiet. I still can't quite believe it. I woke up this morning with the trailer full of strange sounds; little creakings, a humming under the kitchen sink, a strange sporadic clatter from the electric furnace. It took me a few minutes to understand it. When I did I went and stood in the doorway and looked at the plant. The day was cold—it is now spring again, the spring of 1978—but there was no steam from the stacks, and just a kind of wisp rising from the cooling towers. I know what it is, of course: Occasionally maintenance requires that a unit be shut down temporarily, and coincidence has quieted both plants on the same day. Nevertheless, it is eerie here.

Oddly, the quiet seems to make the air clearer; the whole valley seems more sharply defined than I have ever seen it; colors are enhanced, the sky is richer, the coal pile near the plant is more profoundly black. It's like the difference between an enlargement of a thirty-five-millimeter negative and an eight- by ten-inch negative; all the edges of things seem etched, sharpened. Perhaps the constant noise has its effect on the eyes, too. I glance quickly out at the dragline on the horizon, wondering if maybe I am the one stopped, suspended in time, but no, the dragline is still swinging away, dumping its load of overburden in a distant slow-motion crash.

There should be conclusions here. There should be an ending—to this book, to this controversy. But there isn't. There are only more stories; only more conflict. The fate of Colstrip Three and Four is still unknown. It has been five years since the five companies first applied to build it on the Montana prairie. It goes on; it goes on. The power plant never gets closer, and it never gets farther away; it is permanently poised on the brink, but on the brink of what no one knows. It is like a sandstone boulder balanced on a pinnacle, its supports eroding a little

293

more each day—if you had pried at it with a crowbar yesterday it would not have fallen, and dynamite wouldn't have dislodged it last year, but soon it will topple—this week or next century—and on that day the way it falls will only be determined by the direction of the wind. And perhaps by then that's all the ammunition either side will have left.

This year the focus is on the Northern Cheyennes. It is at last 1978, the centennial of their great walk back. Now they are marking it with a full flex of their power. Can they stop the power plant? The achievement of Class I air was just a start. All during the fall of 1977 the Environmental Protection Agency repeatedly stated that it could not grant Colstrip Three and Four a permit because the plant would violate the Class I air; at the beginning of this year it announced it was preparing to change its mind, to squirt Three and Four through a loop-hole in the Clean Air Act amendments of 1977. The spider stalked the reservation. The Northern Cheyennes spent more money on attorneys, persuaded air-quality consultants to testify at a vicious hearing held in Billings, and waited. As I sit here looking out my window at the still plant, the Indians remain waiting, and in my imagination the power plant also waits, for another in the endless series of verdicts and decisions that never seem to create a clear beginning or an end.

But I shall not wait for this conclusion. I'm leaving. I get up from watching the plant and start to pack my belongings. I am selling my trailer and getting out of Colstrip. I throw canned soup in a cardboard box. I put pots and pans in a suitcase, then, choosing between them and books, decide to leave the pans for the next boomer that comes along. I strip the bed. I mop the linoleum. I unplug the refrigerator and open its door. My footsteps sound hollow on the floor; everything I do seems to clank, as if I was listening to my own ears in a circular amplification. I peel the plastic sheets off the inside of the windows; after a year and a half they are stiff with age. They crackle and split as I throw them in the garbage.

Tomorrow morning I will be on the Interstate headed west, past Billings and Livingston and Bozeman, where Marie Sanchez met Russell Means; over the continental divide and into Idaho. In two days I'll be through there, into Utah, where Utah Power and Light is building new coal-fired power plants on the east of the Wasatch Mountains, and where the ghost of the 3,000-megawatt Kaiparowitz power plant, shot once at high noon on the red sandstone plateaus of southern Utah, stalks that array of national parks with a leer that says these things never die. In three days I will be in southern California, the imperial consumer, whose water lines reach out for hundreds of miles, whose power plants stand in Arizona and Nevada, blowing their smoke in other people's eyes.

On a warm winter evening I will sit in a country club in Santa Barbara, California, a feast from a salad buffet in front of me on the cloth tablecloth, trying to answer questions about the book someone has heard I am writing.

It is bright in the room, lit by chandeliers; the desserts here, I have been told, are too sweet. I am not ready to pronounce a verdict on what I have seen, heard, in that state of Montana that seems so far away, but it is required. How can you write a book and be unable or unwilling to produce its theme, like an hors d'oeuvre, at a social gathering? So I stumble along. What is it they want to hear?

I wrote a story about a power plant, I say. A story about a power plant that became a symbol, to me, of the oppression of technology, that the point has been passed at which technology that is intended to enhance human beings now controls, diminishes, enslaves them. I pause. How trite, I think, how convenient. Faces flash before me, over the shrimp salad. Don Bailey, Martin White, Sheila McRae, Wally McRae, Lois Olmstead. I feel like a panelist on a television show, shearing through huge thickets of reality to find one stalk of conveniently vigorous opinion. Vic Jungers, Myron Brien, Clifford Powell, Pat Hayworth, go down before my blade; the power plant, its noise, its voice.

But the people I am sitting with accept what I say. They swallow it. Alternatives, they ask.

Decentralize, I say with vigor. Return the generation of power to communities that use it, not just, or even primarily because they're being imperialistic in planting costs in someone else's garden, but for their own health and integrity, their own sense of personal control. Diversify sources of heat, light, motive power; bring the people back in touch with the reality of their consumption.

These aren't even my ideas at all, I'm thinking, just rote kind of stuff that I've picked up along the way and that has little place in my book but is most convenient to produce at moments like these. Control, diminish, humanize the technology before it completely gets away from us and we are forced to sacrifice fundamental rights, as we are now sacrificing valuable assets—natural beauty, human diversity, respect for land, economic stability—to the ever-increasing demands of its maintenance. I pause, steeped in my own pomposity.

But aren't these plants more efficient? comes a voice across the table.

I'm ready for that one, another little piece of conventional energy debate. No, not at all. More cost-efficient in terms of dollars, perhaps, but in an energy-based economy, terribly inefficient. Two-thirds of the heat is wasted; makes clouds—

But, says a man in the group, if you have the choice of giving lung cancer to one hundred thousand people in a city or giving it to a few people in the country, isn't that a decision that has to be made?

Lung cancer? I think. Lung cancer! Hypothetical arguments run to strange terrors. What right, I say, do we have to inflict lung cancer on a minority so that we can enjoy the benefits free from their costs? Lung cancer? I have a terrible temptation to quote Interior Secretary Carl Schurz' 1878 homily about historical principle and equal rights. But fortunately for all of us I can't quite remember the phrasing—I would stumble as badly as Bill Parker. Crazy arguments, it's not lung cancer we're talking about at all. Something else, something else.

I don't mean we should give it to those people, the man says. If it's going to be that kind of a problem, move them. We've made that kind of decision as a nation before. The majority rules in this nation.

That's what I was thinking about. Not lung cancer. I almost laugh. There is no humor in his face; I can't tell if he's being a devil's advocate or expressing genuine belief.

Move them? I ask. Move them where? Knowing the answer that will come.

Move them to the city.

I smile. I think: What would they say if I told them my book wasn't really about energy at all?

The conversation drifts to the choice between rib roast and chicken Kiev. I retreat into memory.

67

Early in 1977 Wally McRae was invited to speak at the University of Montana, in Missoula, clear on the other side of the state. He was the guest speaker under the auspices of the university's Program Council, which had brought speakers such as William O. Douglas and John Dean to the campus. He spoke in the University Ballroom on a stage; he looked small and slender alone up there, hatless, holding a microphone in his hand with only his boots and his pale forehead to tell he was a rancher.

"The Missoula gig," he wrote later, "was one of the most eerie, and confusing speaking engagements I have ever had." When he stood up on the stage and looked out into the audience, it was gone. The lights that glared in his face had swallowed it; all there was out there was darkness. There were flickers where glasses turned and reflected the light, but the faces were shadowed and he couldn't play the crowd. So he began his speech, the concepts and arguments cracking on each other like rocks pouring out of a dump truck. Impact. Crime. Drugs. Subsidy of industry. Pollution. Energy. Coal.

He paused. There was nothing. The audience was a void, breathing, behind the curtain of light that blinded his eyes. Speaking to them was like throwing those rocks into a canyon and never knowing where they hit. He knew his power, but he was out of reach of it himself, he was so used to it. In other places it had moved people—he had seen it happen, the strength of his voice, his image, his eloquence, had brushed against him, but he had long ago lost the reality of it. It was like a great classic play, rehearsed for months until all the nuances were second nature and the power of it sent the patrons reeling out into the night lost in majesty but left the player only drained and weary, and slightly disgusted with his own art.

He rolled out the speech. Environmentalists. Eminent domain.

Water. Aquifers. People. Megawatts. Indians. Power plants. Who was this silent crowd, huddled around his fire? What of his fears could they possibly understand, four hundred miles from Colstrip? He hurled out the words. Conservation. Limits. Industrialization. There was a slight rustling on the edges of the room, a breeze of life, but he couldn't read it.

"I could see the doors and I could see nobody was leaving," he recalled. "So I thought, well, maybe I'm safe. So I decided, by God, I have to quit somewhere here, so I read 'em this poem that I wrote." The poem came out hard and fast, as it always does when he reads:

> *A smoking limousine darkly does careen*
> *Speeding, ever speeding through the night,*
> *Driven by a child, the car is running wild*
> *The riders all are drunk with giddy fright.*
>
> *Shrieking down the road, with its frenzied load.*
> *Faster, ever faster, on its way.*
> *Dotted lines that merge. Writhing landscapes surge.*
> *The headlights now are off, as on they sway.*
>
> *Speed's narcotic thrills; its enticing siren thrills*
> *It's downhill now, the babbling voices say.*
> *Change is the driver's name. Progress is his game.*
> *The road signs all say "Growth" along the way.*

McRae stopped. It was as if he had turned off a machine setting rivets. There was no rustling now on the edges of the crowd. Except for a glimmer here and there they could have all left. McRae had at last been caught up in his words, and now he was swept away by this silence. He didn't know where he was, the town, the week, the day. He just knew this sudden pause was filled with something strong between him and five hundred people sitting in the dark.

"Thank you," he said.

The house lights came up in the quiet like the glow of the land returning to a ship that has been away. And then the people started to applaud. The noise rose; it boomed; and suddenly he saw that they were standing.

Months later he sat in his chair at the head of his kitchen table at home in the pit of a long night, and let me read a letter he had written to a reporter who had covered the speech. We were stuck somewhere between the Northern Cheyennes' clean air and the County Commissioners' bonds, and McRae was very tired. I had thought he was close to giving up, locked in despair. "If they get Three and Four," he had said earlier in the evening, "it isn't the end, it's the beginning. I don't think we'll be able to make that kind of effort again. We'll cope with what happens with cynicism." The room was quiet as I read the letter.

"The battle has been so long, tough, and trying that I sometimes wonder if I have any remaining philosophy," McRae had written at the end. "My hope is that the five hundred people who gave the standing ovation in Missoula were applauding a philosophy, from a man who possibly has none." They were the words of a man who had already lost what he was fighting to save.

"That's about my bottom line," McRae said. He stared down the length of the table. He gave a long sigh. Then his narrow eyes flicked over at me and then back to deadpan. "No, it isn't," he said. "I don't want people to know my bottom line."

68

On a Saturday in the fall I went bowhunting with Martin White. He carried the bow; I a camera. We went up into the Sarpys, the highest hills in Rosebud County, where the trees are tall and slender and the underbrush thick.

We left his truck parked near the end of a dirt road and hiked silently up into the hills. It had rained the day before, and everything was wet, glittering in the soft early light. The clouds were low, slowly breaking under the gaze of the sun. They raced along from the tops of the hills and occasionally there were gusts of sunlight through the holes. The light flashed off the low leaves of the brush, and the meadows gleamed like streams running between the trees. The pine needles seemed to be made of tarnished silver.

There were hard wild plums in the draws, red as crab apples; we ate them and the black chokecherries and the little dusty-blue Oregon grapes hidden in bunches underfoot, all with a wildness to their flavor that was dry and ripe and bitter all at once. We walked through a field of silver sage, pale and brilliant and wet against our legs, and we separated to explore the two sides of a draw. We never talked above a whisper; our footsteps were the loudest sound we heard for three hours.

At the top of the draw there was a flat space; trees and a meadow. The sun was stronger now, and the Ponderosas steamed. In the clearing we saw three deer that snorted and bounded away, gray shapes in distant forest haze. On a steep hillside White finally got a shot. We had caught a pair of bucks grazing, too intent on morning feed to hear our approach. I watched the deer as White fit an arrow and let it fly. It whirred, missed, and smacked a tree in passing with a sound no more threatening than a twig snapping, and White was delighted, even though he had missed.

"We caught those old guys just sleeping in there," he said. "Not often you can get within fifty yards of a couple of old smart bucks like that."

We came to a ridge and a doe broke down it and into the draw below. We didn't follow. We sat, instead, on the break in the hillside and looked out on the valley. The clouds were layered now, some below us, some above, and all in pieces; so we seemed to be alone here in a separate world. The deer snuffled and crackled along somewhere down below, in dense brush, and we broke our silence with an apple apiece, the crunching sounding horribly loud after hours of whispers.

For a long time we sat with our separate thoughts, and I wondered what White's were. In our last interview he had told me about a dream he nurtured for his retirement. It was an elaborate plan, which meshed with his hunting, his concepts of energy use, and his love of the land. He said he wants to go up to the Big Hole country of western Montana, where the Wise River flows down from the high meadow town of Wisdom, and buy some land on the river, maybe one hundred acres, "just enough to maybe have a horse in case I get old enough I don't like to hike all over those hills." And he said he wants to build a modified A-frame house with sod on one side and a solar collector on the other, a windmill in the back, a greenhouse on the lower floor, and a sewage composter under the floor. "If you could get a setup like this I think it would be a feasible thing that a person could live virtually out of their house and off the land. I wouldn't let anybody else touch building it. I wouldn't let anybody else drive one spike. I'd just love for Sheila and I to do that together."

But we didn't talk about it, sitting on the hill. White was looking down into the valley. He launched his apple core into the draw; mine followed. But we remained there in the peace of the morning.

The clouds slowly withdrew, revealing low brown ridges of ranch land, a tiny reservoir, a dirt road, and the power plant. Suddenly it was there at the edge of the overcast, the twin steam pillars from the stacks looking like supports for a roof of cloud. The clouds seemed thicker downwind of the plant, and White remarked that he ought to send Wally a bill for rainmaking.

We sat quietly. The deer had stopped her travel in the draw. Then White said abruptly:

"I can hear the plants."

And now that he pointed it out, so could I. It was a very distant sound, a deeper mumbling than it was when I stood in its shadow. It was a fundamental sort of noise, as if it was resonating in the earth as well as the air; it was a deep, penetrating undercurrent of sound. It was a hum, a rumble, and it never changed key.

White looked at the steam, listened some more, then said:

"Those plants are too loud."

And I couldn't help remembering, if he did not, his father and him sitting beside the carcass of the biggest elk he had ever killed, listening to the sound of the jeep climbing through the snow in the next canyon.

Epilogue

In 1978 the Environment Protection Agency once again denied the request of five companies for a permit to build Colstrip Units Three and Four as designed, saying the plants would unduly pollute the Northern Cheyenne Indian Reservation. In July 1979, after the second peak of the decade's energy crisis, the agency reversed itself. The United States Department of Energy released a report suggesting that ten Montana counties, including Rosebud, could feed and support thirty-six synthetic fuel plants, each probably close to the size of Colstrip Three and Four. In March 1980 it seemed certain that a bill creating an Energy Mobilization Board would become law. The Board would have the power to force plants such as Colstrip Three and Four around and through legal constraints such as the Montana Major Facility Siting Act, which people like Wally McRae had struggled to develop during the previous years. The board was a miniature energy supremacy act; more may come.

In the spring of 1980 construction began on Colstrip Three and Four, although slowly. One or two small legal details still impede it. But the boomers are coming back. Clint, Natalie, and Allison McRae notice new kids coming to school nearly every week.

Martin White has left Colstrip and moved to Butte; there are rumors he will be named president of Western Energy, replacing Paul Schmechel, who is now president of Montana Power Company. The Olmsteads are back in Colstrip. Vic Jungers has moved to Forsyth. Myron Brien has left the state. Brother Ted Cramer has left the state. Bill Parker is back in Lame Deer. Sheila McRae married Mike and changed her name. They live in a forty-foot-long fifth-wheel camper in the Burtco Court.

303

Two nights before they left, Martin and Sheila White had dinner down on the Rosebud with Wally and Ruth McRae.

The winter of 1978–79 was the toughest in memory for the ranchers who live on Rosebud Creek. They survived.